Lecture Notes in Physics

Edited by J. Ehlers, München, K. Hepp, Zürich, R. Kippenhahn, München, H. A. Weidenmüller, Heidelberg, and J. Zittartz, Köln
Managing Editor: W. Beiglböck, Heidelberg

70

Wave Propagation and Underwater Acoustics

Edited by
Joseph B. Keller and John S. Papadakis

Springer-Verlag
Berlin Heidelberg New York 1977

Editors

Joseph B. Keller
Courant Institute
of Mathematical Sciences
New York University
New York, New York/USA

John S. Papadakis
New London Laboratory
Naval Underwater Systems Center
New London, Connecticut/USA

ISBN 3-540-08527-0 Springer-Verlag Berlin Heidelberg New York
ISBN 0-387-08527-0 Springer-Verlag New York Heidelberg Berlin

Printed in Germany

Printing and binding: Beltz Offsetdruck, Hemsbach/Bergstr.
2153/3140-543210

£6.20

WAVE PROPAGATION AND UNDERWATER ACOUSTICS

Joseph B. Keller* and John S. Papadakis,** Editors

* Courant Institute of Mathematical Sciences, New York University

** Department of Mathematics, University of Rhode Island and
Naval Underwater Systems Center, New London Laboratory

Preface

A "Workshop on Wave Propagation and Underwater Acoustics" was held from November 19 to November 21, 1974 in Mystic, Connecticut. It was sponsored by the Acoustics Branch of the Office of Naval Research under the aegis of Hugo Bezdek. The workshop was conceived at the New London Laboratory of the Naval Underwater Systems Center and organized by the following committee of members of that laboratory:

Chairman: John S. Papadakis, Department of Mathematics,
University of Rhode Island (Consultant)

L. T. Einstein
R. H. Mellen
Henry Weinberg

Among the twenty-one lectures at the workshop was a set of six surveys of various aspects of the field. Those surveys were presented by five members and one former visiting member of the Courant Institute of Mathematical Sciences, New York University. They were prepared with the intention that they would be expanded, combined and published together as a general survey of the mathematical theory of underwater sound propagation. These notes are the result. They would not have appeared without the untiring effort of Professor John S. Papadakis, who guided them through the editorial process. I wish to thank him particularly for this. I also thank the entire committee for having asked me to present a set of survey lectures, and for then agreeing to let me share the presentation with my colleagues.

Joseph B. Keller

WAVE PROPAGATION AND UNDERWATER ACOUSTICS

TABLE OF CONTENTS

CHAPTER I

SURVEY OF WAVE PROPAGATION AND UNDERWATER ACOUSTICS

Joseph B. Keller

Courant Institute of Mathematical Sciences
New York University
251 Mercer Street
New York, NY 10012

1. Introduction

Underwater acoustics, the science of sound propagation in the ocean, has been developed extensively during the last forty years in response to practical needs. By now the theory is so well developed that it provides a general understanding and a detailed description of how sound travels in the ocean, and of the mechanisms affecting it. The theory can also be used to make quantitative calculations of the sound field produced by a given source. However, there are difficulties which limit the accuracy of such calculations. The first is the lack of adequate information about the sound velocity in the ocean as a function of position and time. The second is the analytical and computational difficulty of calculating the sound field in terms of the properties of the ocean. The mathematical methods which have been devised to overcome this latter difficulty are the subject of these notes.

The analysis of underwater sound propagation is based upon the physical principles of theoretical acoustics. These principles lead to a wave equation for the acoustic pressure, together with suitable boundary conditions at the ocean surface and bottom, and initial conditions. The properties of the ocean which enter into this formulation are the sound speed $c(x,y,z,t)$, the bottom depth $h(x,y)$, the surface elevation $\eta(x,y,t)$ and the ambient water velocity $\underline{u}(x,y,z,t)$. Absorption, which results from viscous dissipation, heat conduction, chemical reaction, scattering by particulate matter, etc. is usually accounted for by an absorption coefficient which depends upon position and frequency. In the analysis of time harmonic fields it is combined with the sound speed to yield a complex refractive index. Absorption by the bottom is usually accounted for by a bottom impedance, or sometimes by a bottom reflection coefficient.

Most of the theoretical analyses ignore the surface elevation, the ambient water velocity, and the absorption in the fluid and in the bottom. Some of these effects are taken into account afterwards in an ad hoc manner. For the most part we shall follow the common procedure of ignoring them.

Initially, the theory concerned the deterministic problem of propagation in an ocean of prescribed constant or gradually varying properties. However, as experimental technique improved, it was found that the observed sound field undergoes extensive and rapid fluctuations. These fluctuations are caused by fluctuations in the properties of the ocean. To analyze them the ocean is represented as a random medium, and the problem of sound propagation in a random medium is considered. The theory of this kind of propagation is not as well developed as that of propagation in a deterministic medium, as we shall see. We shall first describe the theory of the deterministic case and then describe that of the random case.

2. Wave propagation in a deterministic medium

Let us consider first the simplest case, that of a time harmonic point source in an unbounded homogeneous ocean. The resulting sound field is a spherical wave. Secondly, suppose the ocean is bounded above by a horizontal plane free surface on which the acoustic pressure p vanishes. Then p is the sum of two spherical waves, one from the source and another from the image of the source in the plane surface, multiplied by the reflection coefficient $R = -1$. The interference between these two waves leads to an oscillation in the magnitude of p which is sometimes referred to as the Lloyd mirror effect. Thirdly, let the ocean be bounded above by a horizontal plane free surface on which $p = 0$ and below by a horizontal plane bottom on which the normal derivative $\partial p/\partial n = 0$. Then p is the sum of an infinite number of spherical waves from the source and from an infinite set of images of it in the two planes.

The image method of constructing p, which leads to the above results, does not generalize to the case of an inhomogeneous ocean nor to the case of non-planar boundaries. Furthermore, at horizontal distances from the source which are large compared with the depth, many of the spherical waves have nearly the same phase, or arrival time. This makes it difficult to calculate p because the successive waves nearly cancel one another.

These disadvantages of the image method can be overcome, in part, by the method of normal modes, which was introduced and developed by C.L. Pekeris [1]. That method applies to any horizontally stratified ocean of constant depth. It leads to a representation of p as the sum of an infinite number of normal modes. Only a finite number of them are propagating and the rest are evanescent. Thus, at large distances from the source only the propagating modes are important, so there p is represented by a finite sum.

The method of normal modes is restricted to horizontally stratified oceans of constant depth. Furthermore, at distances from the source which are not large compared with the depth, the evanescent waves are not negligible, so many of them must be taken into account in calculating p.

The latter difficulty, but not the former, can be overcome by the Hankel transform method, which was utilized by L. Brekovskikh [2] and others. This yields a representation of p as an integral involving Bessel functions and solutions of the normal mode equation. Although this integral is convenient for evaluation at short ranges, it is not so convenient at long ranges, where the normal mode representation is more useful.

A third representation of p in a horizontally stratified ocean of constant depth is given by the method of multiple scattering. This method is a generalization of the image method from the case of a homogeneous ocean to that of a horizontally stratified one. In it p is represented as a sum of waves: one wave emerging directly from the source, another wave which represents scattering of the direct wave by the medium above the source, a third wave which results from scattering of the direct wave by the medium below the source, and successive multiply scattered waves. Scattering includes both reflection by a boundary and refraction by the inhomogeneous medium. In the case of a homogeneous ocean, refraction is absent and scattering is only reflection. In this case, the multiple scattering representation reduces exactly to that given by the image method.

Since the three representations of p described above are all exact, they all yield the same value of p. Furthermore, each representation can be converted into either of the other two. In addition, each representation can be simplified by conversion into an asymptotic form which is valid when the acoustic wavelength is small compared to the distance over which the sound speed varies appreciably. The asymptotic forms can also be converted into one another.

The asymptotic form of the multiple scattering representation has an interpretation in terms of the rays of geometrical acoustics. The asymptotic form of an n times scattered wave is just the geometrical acoustics field on a ray which has been reflected and/or refracted n times. Therefore, this asymptotic form is called the ray representation. It has two important advantages over the other representations, which we shall now describe. The first is that it provides a geometrical and physical picture of how propagation occurs, and it shows where the sound goes. The second advantage results from the fact that the ray representation can be derived directly without the restriction to a horizontally stratified ocean of constant depth. Therefore, a ray representation can be obtained for a general ocean with horizontal as well as vertical variation of sound speed, and with depth variation.

A more refined asymptotic analysis also yields surface diffracted rays. These rays are produced at the ocean surface and bottom by refracted rays which are tangent to those surfaces. They travel along the surface or bottom within the ocean and refract back into the interior. In addition, if propagation within the bottom is considered, and if it is faster than that in the ocean, the asymptotic analysis yields a head wave. It is associated with rays which hit the bottom at the critical angle, travel in the bottom along the interface, and re-enter the ocean at the critical angle. We shall not consider these effects nor shall we consider the consequences of using an impedance boundary condition on the bottom.

The ray representation also has two disadvantages. The first is that it becomes infinite on the caustic surfaces of the rays, and is invalid there. Consequently, a different representation, such as a boundary layer expansion employing Airy functions, must be used on and near each caustic. Alternatively the uniform representation, introduced by D. Ludwig [3] and by Yu. A. Kravtsov [4], can be used both near and away from each caustic. The second disadvantage is that it is difficult to evaluate numerically the expression for the amplitude on a ray in the general case.

A second method for taking account of horizontal and temporal variations in sound speed and bottom depth employs a combination of normal modes and horizontal or two dimensional rays. Each normal mode is assumed to propagate independently of the others. Its horizontal velocity at each point x,y on the surface is determined by the vertical sound speed profile and the depth beneath that point. This horizontal velocity is used in the construction of horizontal rays, which determine where each normal mode travels. The amplitude of each mode is determined by a transport equation along each horizontal ray.

The construction of the sound field by this method proceeds as follows. First, the vertical structure and horizontal velocity of each normal mode must be found at each point x,y. Second, for each mode, the horizontal rays which start from the point above the source must be found. Third, the initial amplitude of each normal mode on each ray must be determined from the source strength distribution. Fourth, the phase and amplitude of each normal mode must be found at each point on each horizontal ray, starting with the values at the point above the source. Fifth, beneath any point x,y the sound field is given by the sum of the normal mode functions at that point, each with the phase and amplitude determined from the corresponding ray from the source to x,y.

This method can be derived systematically from the assumptions that the horizontal and temporal gradients of sound speed and bottom depth are small. The derivation also yields corrections to the theory if the gradients are not so small. This type of theory and its derivation were introduced in 1958 by J. B. Keller [5] in the analysis of surface waves in water of nonuniform depth. For underwater sound it was introduced by A. D. Pierce [6] in 1965. The systematic derivation of the theory for this case, together with its implementation and application, were presented by H. Weinberg and R. Burridge [7] in 1974.

The method of normal modes and horizontal rays enjoys some advantages of each of the two methods which it combines, and avoids some of their disadvantages. Thus, it is applicable to oceans with horizontal and temporal variations in sound speed and depth, whereas the normal mode method is not. However, the horizontal and temporal gradients in these quantities must be smaller than is required for the ray method alone. It avoids the necessity of finding rays in three dimensions and constructing the amplitudes along them. But it still fails to be valid at the caustics, which are now curves in the horizontal plane, and on the vertical lines through the caustics. Again boundary layer expansions and uniform expansions can be used on and near these places.

An alternative to the use of horizontal rays together with normal modes is the use of a horizontal wave equation for the complex amplitude of each normal mode at x,y. This theory is sometimes called NINMA, an acronym for "non-interacting normal mode analysis". It also can be derived when the horizontal gradients of sound speed and bottom depth are small. Its advantage over the use of normal modes and horizontal rays is that it avoids the non-uniformities associated with caustics. Its disadvantages are that it requires more computing, since one must solve a wave equation for the amplitude of each mode, and it does not provide the geometrical picture of where the modes travel, which is provided by the horizontal rays. NINMA is not described further in these notes.

A third method for dealing with horizontal variations in sound speed and depth is the parabolic equation method. This is a method for the approximate description of time harmonic waves which are propagating primarily in one direction. It was introduced in connection with electromagnetic wave propagation by M. Leontovich and V. A. Fock [8] in 1946 and adapted to underwater sound propagation by F. D. Tappert and R. H. Hardin [9] in 1973. To illustrate the method we consider the equation

1. $$p_{xx} + p_{yy} + p_{zz} + k^2 n^2(\underline{x}) p = 0$$

For a wave travelling primarily in the x direction we write $p = e^{ikx}q$, and we find that q satisfies the equation

$$2. \qquad q_{xx} + 2ikq_x + q_{yy} + q_{zz} + k^2[n^2(\underline{x})-1]q = 0 .$$

We now assume that q_{xx} is small compared to $2ikq_x$, so we drop it and obtain

$$3. \qquad 2ikq_x + q_{yy} + q_{zz} + k^2[n^2(\underline{x})-1]q = 0.$$

This equation for q is of first order in x and of second order in y and z , like a parabolic equation with x playing the role of time.

The advantage of (3) over (1) is that (3) is easier to solve numerically. This is because the parabolic character of (3) permits it to be solved by a marching or step-by-step method in the x-direction. In contrast (1), being elliptic, must be solved simultaneously at all values of x. This difference makes it possible to solve problems involving (3) which it would be impossible to solve, or very difficult to solve, using (1). In the application to underwater acoustics, the role of x is played by the cylindrical coordinate r and e^{ikx} is replaced by the Hankel function $H_0^{(1)}(kr)$.

In order to use (3) or the corresponding equation with r instead of x , it is necessary to derive initial conditions at $x = 0$, or $r = 0$. These conditions are obtained by matching the solution of (3) to the solution of (1) near the source.

The practical value of (3) can be greatly improved by using the efficient numerical methods which are available for the solution of parabolic equations. Tappert has refined these methods to the point where it is possible to solve (3) repeatedly with different random choices of $n(\underline{x})$ to simulate wave propagation in a random medium.

The disadvantage of the parabolic equation method is its limitation to nearly radial propagation. It is inaccurate whenever the rays from the source deviate appreciably from horizontal straight lines. Therefore, it is not valid if the rays bend significantly in either a horizontal or a vertical plane, or if the bottom slope becomes large. However, within these limitations it appears to be very useful.

3. Wave propagation in a stochastic medium

The observed temporal fluctuations in the sound field, mentioned in the Introduction, may be due to temporal fluctuations of the sound speed in the ocean, to temporal fluctuations in the ambient velocity of the water, and to temporal fluctuations in the elevation of the ocean surface. To analyze these fluctuations it is customary to consider each temporally fluctuating quantity to be a random quantity. Then the acoustic pressure is the solution of a wave equation in which some coefficients are random functions, and in which the upper boundary is a random surface.

This treatment of the pressure fluctuations in terms of a stochastic differential equation with a stochastic boundary raises two problems. The first concerns the relationship between the solution of this stochastic problem and the observed pressure. The second problem is the mathematical one of solving the stochastic problem.

The first problem is usually resolved by tacitly assuming that the statistical properties of the theoretical pressure will agree with the corresponding statistics of the observed pressure provided that the statistical properties of the ocean are chosen properly. In this statement the theoretical problem involves a stochastic pressure and a stochastic ocean, while the statistics of the observed pressure are based upon a temporal record and time averaging. Therefore the statistical properties of the theoretical stochastic ocean should agree with the temporal statistics of the actual ocean in order to be appropriate. This necessitates the observation of the statistics of the sound speed,of the ambient velocity of the water, and of the ocean surface. Considerable progress has been made in this direction, but much more remains to be done. We shall not consider this problem further.

The second problem is a special case of the general one of wave propagation in a random medium, with the extra complication of a random boundary. There are several reviews of this subject, such as the books of Chernov [10], of Tatarski [11], and of Klyatskin [12], the articles of Keller [13], of Frisch [14], of Barabanenkov , Kravtsov, Rytov and Tatarski [15], and those in the book edited by Keller and McKean [16]. Other relevant works are referred to in Chapters IV and V. Therefore we shall present only a brief description of this extensive field. Chapter IV contains a detailed investigation of a special aspect of it, namely the analysis of the stochastic equations for the amplitudes of the normal modes of a sound field in an ocean with a random sound speed.

In principle the stochastic problem can be formulated by introducing a family of oceans depending upon a random variable α , with a probability density $P(\alpha)$. The solution of the propagation problem for each α yields a pressure $p(\underline{x},\alpha)$ which is also random, since it depends upon α. Then the statistics of $p(\underline{x},\alpha)$ such as its mean, its variance, its two point correlation function, etc. can be calculated using the solution $p(\underline{x},\alpha)$ and the probability density $P(\alpha)$. This procedure can be described as "solving and then averaging" to get the statistics of p.

A common method of solving for p is the Born expansion, an expansion in powers of the deviation of the sound speed from a constant. This expansion has the defect that any finite number of terms of it leads to divergent results in a statistically homogeneous medium of infinite extent. Nevertheless Pekeris (see[10]) showed how to use the first Born approximation to calculate the average energy scattered from a wave by unit volume of the medium. This determines the total scattering cross section per unit volume and the corresponding attenuation coefficient. The latter can be used to calculate the exponential decay of a propagating wave. His method also gives the average energy scattered into any direction by unit volume of the medium, which yields the differential scattering cross section per unit volume. The differential cross section can be used in the transport equation for the incoherent energy flux in the medium.

The second Born approximation was applied to a thin slab of the medium by Keller [13]to determine a modified propagation constant. This constant governs the propagation of the average wave. Its imaginary part is just the attenuation coefficient obtained by Pekeris. Originally Rayleigh had used the slab method to find the modified propagation constant in a medium containing discrete scatterers, such as dust particles or water droplets.

To overcome the defects of the Born expansion, various methods have been employed. One of the simplest and most useful is the forward scattering approximation to the first Born approximation. In this approximation only scattering into the "forward" half-space is taken into account, so the first Born approximation reduces to an integral over the region between the source and the observation point. This simplification leads to finite results, but the scattered intensity increases indefinitely as the observation point moves away from the source. Thus the forward scattering - Born approximation is not uniformly valid with respect to the position of the observation point.

The lack of uniformity of the forward scattering - Born approximation is partly overcome by the Rytov method. This is a modification of the Born method in which log p is expanded rather than p. It leads to the same integrals as the Born expansion, so it also yields divergent results in statistically homogeneous media of infinite extent. When modified to the forward scattering - Rytov approximation, however, it gives finite results. They are not uniformly valid with respect to the observation point either, but they are valid to a much greater range than those of the forward scattering - Born expansion.

Another method of avoiding the divergence of the Born approximation is that of summation of a selected infinite subset of terms in the Born expansion. This can be done for the average of p and for the two point correlation of p, assuming that the sound speed fluctuations are Gaussian. The terms are usually represented by Feynman diagrams. Then it is shown by diagrammatic means that the sum of the selected terms satisfies a certain integral equation or integro-differential equation. These equations are analogous to the Dyson and Bethe-Salpeter equations of quantum field theory. The introduction of these equations for the average and two point correlation of p is often called the smoothing method because these equations have smooth coefficients.

A much simpler derivation of the smoothing method for the average field was given by Ament for electromagnetic waves, by Meecham for scalar waves, and by Bourret for general waves. (See [13]). They averaged the original equation and replaced a certain average of a product by a product of averages. A direct perturbation theoretic derivation of this result, for general waves, and of that for the two point correlation function, was given by Tatarski and Gertsenstein (See [14]), and by Keller [13].

The equation for the average pressure can be solved to yield a modified propagation constant which is essentially the same as that obtained from the second

In view of the importance of this model, we shall analyze it in some detail. First we shall obtain the exact solution for the acoustic pressure p by three well known methods, which lead to three different representations of p. These are the method of normal modes, the method of Hankel transformation and the method of multiple scattering. Then we shall show how these three different representations can be transformed into one another by using contour integration and residue evaluation, the binomial expansion and the Poisson summation formula.

Next, and most importantly, we shall evaluate each of the three representations asymptotically for the wavelength small compared to the scale length of the sound velocity profile. These evaluations involve three methods of asymptotic analysis: the WKB method for the asymptotic solution of ordinary differential equations, the Langer modification of this method to treat equations with turning points, and the method of stationary phase for the asymptotic evaluation of integrals with rapidly oscillating integrands. The resulting three asymptotic representations are simpler than the exact ones and have clear physical interpretations. Each one is most useful in a particular range of parameters. We shall also show how these asymptotic representations can be converted into one another.

Finally we shall obtain the ray representation of p. This is a representation which involves the rays of geometrical acoustics. First we shall obtain it by further asymptotic evaluation of the asymptotic form of the multiple scattering representation. Then we shall show how to get it by a construction involving rays, phase functions, amplitude functions and other concepts of geometrical acoustics. Thirdly we shall derive it by direct asymptotic solution of the reduced wave equation and the corresponding boundary conditions. The last two derivations have the virtue that they are applicable to an ocean with an arbitrary variation of sound velocity with position, and an arbitrary depth variation. It is merely necessary that the scale lengths of these variations be large compared to the wavelength. However the practical utilization of the ray representation is computationally difficult in cases other than that of a stratified ocean of constant depth.

Much of the work on which this chapter is based is due to C. L. Pekeris and to L. Brekhovskikh. More extensive accounts, together with various applications and additional details, can be found in the books of Brekhovskikh [1], Ewing, Jardetsky and Press [12], Felsen and Marcuvitz [13] and Keller and Lewis [14].

1. Formulation and fundamental equations

The velocity $\underline{u}$, pressure p, mass density ρ and entropy density s in an inviscid, non-heat conducting fluid satisfy the following equations:

1.1 $$\underline{u}_t + (\underline{u}\cdot\nabla)\underline{u} = -\rho^{-1}\nabla p + \underline{g} + \rho^{-1}\varepsilon\underline{f} ,$$

1.2 $$\rho_t + \nabla \cdot (\rho\underline{u}) = 0 ,$$

1.3 $$s_t + \underline{u} \cdot \nabla s = 0 ,$$

1.4 $$p = p(\rho,s) .$$

These are the equations of momentum, mass conservation, adiabatic motion and the equation of state, respectively. In (1.1) $\underline{g}$ is the acceleration of gravity and $\varepsilon\underline{f}$ is another external force per unit volume, which represents an acoustic source. The parameter ε is a measure of the strength of this source.

Let us suppose that the fluid is bounded above by the free surface $z = \eta(x,y,t)$ and below by the rigid surface $z = -h(x,y)$. Let p_o be the constant pressure above the free surface. Then the continuity of pressure across the free surface and the kinematic condition at this surface yield

1.5 $$p[x,y,\eta(x,y,t),t] = p_o \quad \text{on} \quad z = \eta(x,y,t) ,$$

1.6 $$\eta_t + u\eta_x + v\eta_y = w \quad \text{on} \quad z = \eta(x,y,t) .$$

The rigidity of the bottom requires that the normal component of $\underline{u}$ vanish on the bottom:

1.7 $$w + uh_x + vh_y = 0 \quad \text{on } z = -h(x,y) .$$

Here $\underline{u} = (u,v,w)$ and $\underline{x} = (x,y,z)$ with the positive z axis pointing vertically upward.

When $\varepsilon = 0$, a particular solution of (1.1) - (1.7) is $\underline{u} = 0$ and $\eta = 0$ with p, ρ and s depending only upon z. These three functions $p(z)$, $\rho(z)$ and $s(z)$ are related by (1.4) and the z component of (1.1), which yields the hydrostatic equation

1.8 $$p_z = -\rho g \ .$$

In addition (1.5) yields $p(0) = p_o$. Thus one of these three functions, or one additional relation among them, can be prescribed. Then (1.4) and (1.8) yield the remaining two functions. We shall call this solution the basic state.

We now consider a particular solution of (1.1) - (1.7), which naturally depends upon the parameter ε in (1.1). We assume that when $\varepsilon = 0$, this solution reduces to the basic state described above. By differentiating (1.1) - (1.7) with respect to ε and setting $\varepsilon = 0$, we obtain the acoustic equations, which are:

1.9 $$\dot{\underline{u}}_t = -\rho^{-1}\nabla\dot{p} + \rho^{-2}\dot{\rho}\nabla p + \rho^{-1}\underline{f} \ ,$$

1.10 $$\dot{\rho}_t + \nabla \cdot (\rho\dot{\underline{u}}) = 0 \ ,$$

1.11 $$\dot{s}_t + \dot{\underline{u}} \cdot \nabla s = 0 \ ,$$

1.12 $$\dot{p} = p_\rho\dot{\rho} + p_s\dot{s}$$

1.13 $$\dot{p} + \dot{\eta}p_z = 0 \ , \ z = 0 \ ,$$

1.14 $$\dot{\eta}_t = \dot{w} \ , \ z = 0 \ ,$$

1.15 $$\dot{w} + \dot{u}h_x + \dot{v}h_y = 0 \ , \quad z = -h(x,y) \ .$$

Here $\dot{\underline{u}}$, $\dot{p}$, etc. denote derivatives with respect to ε evaluated at $\varepsilon = 0$, while p, ρ, s, etc. denote the basic state. We shall call $\dot{\underline{u}}$, $\dot{p}$, etc. the acoustic quantities.

It is convenient to obtain a single equation and boundary conditions for p by eliminating the other acoustic quantities from these equations. To do so we differentiate (1.10) and (1.11) with respect to t, noting that the basic state is independent of t. We then use (1.9) to eliminate $\dot{\underline{u}}_t$ and (1.8) for p_z, to obtain

1.16 $$\dot{\rho}_{tt} - \Delta\dot{p} - g\dot{\rho}_z = -\nabla \cdot \underline{f}$$

$$1.17 \qquad \dot{s}_{tt} - \rho^{-1}(\dot{p}_z + g\dot{\rho})s_z = -\rho^{-1}s_z f_3 \ .$$

Now we differentiate (1.12) twice with respect to t and use (1.16) and (1.17) in the resulting equation to get

$$1.18 \qquad \dot{p}_{tt} - p_\rho \Delta \dot{p} = -p_\rho \nabla \cdot \underline{f} + \rho^{-1} p_s s_z(\dot{p}_z + g\dot{\rho} - f_3) + g p_\rho \dot{\rho}_z \ .$$

In ocean acoustics, all the terms on the right side of (1.18) except the first one are usually negligible compared to the other terms. When this is the case, (1.18) can be replaced by the wave equation for $\dot{p}$:

$$1.19 \qquad \Delta \dot{p} - \frac{1}{c^2}\dot{p}_{tt} = \nabla \cdot \underline{f} \ .$$

Here $c^2 = p_\rho$ is the sound speed, which depends only upon z because the basic state depends only upon z.

In the boundary condition (1.13) the term $\dot{\eta} p_z$ is equal to $-\rho g \dot{\eta}$ in view of (1.8), and this term is usually negligible compared to $\dot{p}$. Then (1.13) becomes

$$1.20 \qquad \dot{p} = 0 \text{ at } z = 0 \ .$$

Finally we differentiate the bottom boundary condition (1.15) with respect to t and use (1.9) to eliminate $\underline{\dot{u}}_t$. As before, we assume that the term $\rho^{-2}\dot{\rho}\nabla p = -\rho^{-1}\dot{\rho}\underline{g}$ is negligible, and also that $\underline{f} = 0$ at the bottom. Then (1.17) yields

$$1.21 \qquad \dot{p}_z + \dot{p}_x h_x + \dot{p}_y h_y = 0 \quad \text{at } z = -h(x,y) \ .$$

The wave equation (1.19), together with the two boundary conditions (1.20) and (1.21), plus the specification of the initial values of $\dot{p}$ and of $\dot{p}_t$, constitute an initial-boundary value problem for $\dot{p}(\underline{x},t)$. Once $\dot{p}$ is found, the other acoustic quantities can be found from (1.9) - (1.11) and (1.14), provided that their initial values are given. In order to find $\dot{p}$ it is necessary to know the sound speed c(z), the bottom depth h(x,y) and the source distribution $\nabla \cdot \underline{f}(\underline{x},t)$, in addition to the initial values of $\dot{p}$ and $\dot{p}_t$. We shall assume that these quantities are known, and consider the methods of solving the problem for $\dot{p}$.

2. Time harmonic waves

The most important acoustic fields are the time harmonic ones, in which $\dot{p}$ is of the form

2.1 $$\dot{p}(\underline{x},t) = e^{-i\omega t}p(\underline{x}) .$$

Here and hereafter it is to be understood that $\dot{p}$ or any other real quantity is the real part of a complex expression for it, such as that on the right side of (2.1). The complex pressure amplitude $p(\underline{x})$ will be referred to as the pressure for short. It is not to be confused with the pressure in section 1, which is denoted by the same letter, but which will not appear again. In order that (2.1) satisfy (1.19), $\nabla \cdot \underline{f}$ must be of the form

2.2 $$\nabla \cdot \underline{f}(\underline{x},t) = e^{-i\omega t}q(\underline{x}) .$$

When (2.1) and (2.2) hold, then (1.19) - (1.21) become

2.3 $$\Delta p + k^2n^2(z)p = q(\underline{x}) ,$$

2.4 $$p = 0 \quad \text{at} \quad z = 0 ,$$

2.5 $$p_z + p_xh_x + p_yh_y = 0 \quad \text{at } z = -h(x,y) .$$

In (2.3) we have introduced the wavenumber $k = \omega/c_o$ and the refractive index $n = c_o/c(z)$, where c_o is some typical value of the sound speed. We call (2.3) the reduced wave equation or sometimes the Helmholtz equation.

The boundary value problem (2.3) - (2.5) does not determine p uniquely. This is because the homogeneous problem, obtained by setting $q(\underline{x}) = 0$, has solutions which represent waves coming in from infinity. Therefore some additional condition must be imposed in order to eliminate these extraneous waves and determine a unique solution. There are three different methods for doing this, which we shall now describe. The first and physically most appealing, is to solve the initial value problem with the source given by (2.2) and with $\dot{p} = \dot{p}_t = 0$ at $t = 0$. This problem has a unique solution $\dot{p}(\underline{x},t)$, which we expect to approach the form (2.1) as $t \to \infty$. Therefore we define $p(\underline{x})$ by

2.6 $$p(\underline{x}) = \lim_{t\to\infty} e^{i\omega t} \dot{p}(\underline{x},t) .$$

It can be proved that this limit exists and that it satisfies (2.3) - (2.5).

The second method is to replace k by the complex quantity

2.7 $$k = \frac{\omega}{c_o} + i\alpha , \quad \alpha > 0 .$$

The positive constant α represents absorption, and therefore the desired solution of (2.3) - (2.5) will decay to zero at infinite distance from the source region. An incoming wave, however, will be infinitely large at infinity in the direction from which it comes. Therefore the requirement that the solution be bounded at infinity should eliminate incoming waves and pick out a unique solution $p(\underline{x},\alpha)$, which depends upon α. Then as α tends to zero, this solution should tend to a limit. Thus we define $p(\underline{x})$ by

2.8 $$p(\underline{x}) = \lim_{\alpha\to 0} p(\underline{x},\alpha) .$$

It can be proved that $p(\underline{x},\alpha)$ exists and is unique, that this limit exists and that it satisfies (2.3) - (2.5) with $\alpha = 0$. Furthermore it is the same solution as is given by (2.6).

The third method deals directly with (2.3) - (2.5) keeping k real. It is to impose a radiation condition on the solution, which directly eliminates incoming waves and thereby selects a unique solution. The precise form of this condition depends upon the number of space dimensions, the shape and depth of the domain, etc. In the present case it involves the normal modes and eigenvalues of the problem, so we shall not formulate it until we introduce those quantities. It can be proved that this method also yields a unique solution $p(\underline{x})$ which is the same as that given by (2.6) and (2.8). The fact that the limit in (2.6) yields the same solution as the method using the radiation condition is sometimes called the limiting amplitude principle, while the fact that (2.8) yields the same solution is called the limiting absorption principle. Thus any one of these methods can be used to find the desired solution $p(\underline{x})$, which we shall call the radiating or outgoing solution.

There is actual absorption of sound in the ocean due to viscosity, heat conduction and chemical reaction, all of which we have ignored in deriving (2.3). This absorption can be accounted for by writing k in the form (2.7) with $\alpha(\omega)$ a function of frequency determined by the dissipative processes. When this absorption is taken into account, the correct solution $p(\underline{x})$ is selected by the requirement that it be bounded at infinity. Then the radiation condition is not necessary. Furthermore the time dependent equation corresponding to the reduced wave equation (2.3) with k given by (2.7) is not just the wave equation (1.19), but is a more complicated equation or system of equations.

The time harmonic solutions (2.1) can be used in a Fourier integral to synthesize the solution of (1.19) for a source with arbitrary time dependence. This accounts in part for their great importance. Thus suppose that $\nabla \cdot \underline{f}$ has the Fourier representation

2.9 $$\nabla \cdot \underline{f}(\underline{x},t) = \int e^{-i\omega t} q(\underline{x},\omega) d\omega \ .$$

Then if $p(\underline{x},\omega)$ is the outgoing solution of (2.3) - (2.5) with the source $q(\underline{x},\omega)$, the solution of (1.19) - (1.21) is

2.10 $$\dot{p}(\underline{x},t) = \int e^{-i\omega t} p(\underline{x},\omega) d\omega \ .$$

Furthermore, the solution for an arbitrary source distribution $q(\underline{x},\omega)$ can be obtained from the solution for a point source, represented by a delta function. Therefore we shall consider the solution of (2.3) with $q(\underline{x}) = -\delta(\underline{x}-\underline{x}_o)$.

3. The homogeneous ocean of constant depth

3.1 Introduction

The simplest sound velocity profile is the uniform one $c(z) = c_o$, where c_o is a constant. In this case $n(z) = c_o/c(z) = 1$, and for a point source (2.3) becomes

3.1 $$\Delta p + k^2 p = -\delta(z-z_o) \frac{\delta(r)}{2\pi r} \ .$$

Here the source location $\underline{x}_o$ is $r = 0$, $z = z_o$, in cylindrical coordinates. The surface condition is (2.4),

3.2 $$p = 0 \quad \text{at} \quad z = 0 \ .$$

We shall assume that the depth is constant so that h = constant, and (2.5) becomes

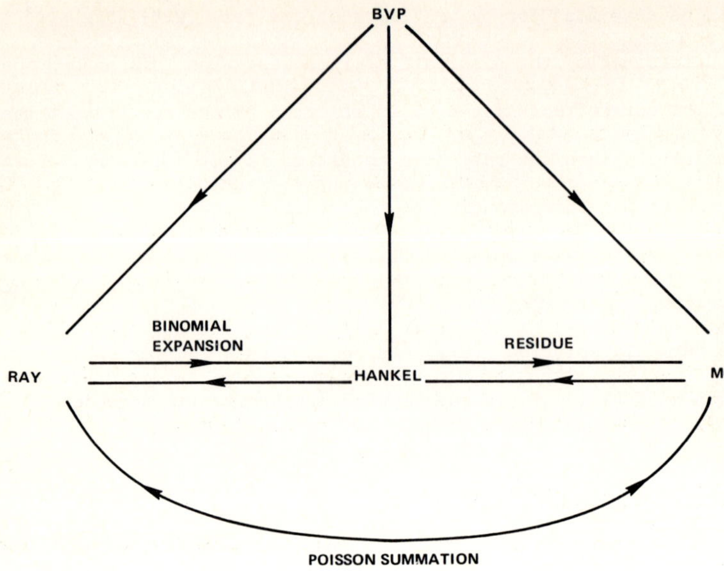

Figure 1. The boundary value problem (BVP) is solved by the method of normal modes, by the method of Hankel transforms and by the ray method. This leads to the three representations denoted by mode, Hankel and ray, respectively. Then the representations are transformed into one another by the method of residues, by the binomial expansion and by the Poisson summation formula, as indicated.

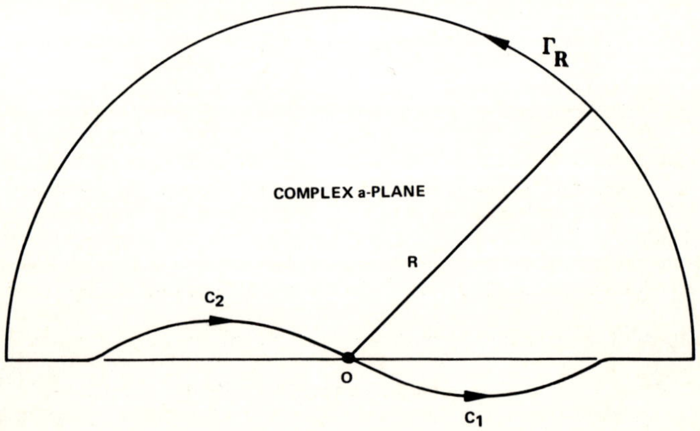

Figure 3. The contour of integration C_1 extends from the origin to infinity and is slightly below the real axis in the a-plane. The contour C_2 is $e^{i\pi}C_1$ with the orientation reversed. The arc Γ_R of radius R connects C_1 and C_2 to form a closed contour.

3.3 $$p_z = 0 \quad \text{at} \quad z = -h .$$

The problem (3.1) - (3.3), with a suitable radiation condition, determines the sound pressure due to a time harmonic point source in a homogeneous ocean of constant depth with a free surface and a rigid bottom. Since the problem is axially symmetric, the solution $p(r,z)$ is independent of the angular coordinate θ.

In the next three sub-sections we shall solve this problem by three different methods and obtain three different representations for the solution. Then in the final sub-section we shall show how these representations can be transformed into one another. All of these results are summarized in Figure 1.

3.2 Normal mode representation

The homogeneous form of (3.1) can be solved by separation of variables. To use this method we seek a solution which is a product $\phi(z)\psi(r)$. We substitute it into the homogeneous form of (3.1) and separate variables to obtain

3.4 $$\phi_{zz} + k^2\phi = k^2a^2\phi ,$$

3.5 $$\psi_{rr} + \frac{1}{r}\psi_r = -k^2a^2\psi .$$

We have written the separation constant as ka for convenience. The general solutions of these two equations are

3.6 $$\phi(z) = A\sin[k(1-a^2)^{1/2}z] + B\cos[k(1-a^2)^{1/2}z] ,$$

3.7 $$\psi(r) = CH_o^{(1)}(kar) + DH_o^{(2)}(kar) .$$

Here $H_o^{(1)}$ and $H_o^{(2)}$ are the Hankel functions of order zero of the first and second kinds, respectively.

The boundary conditions (3.2) and (3.3), when applied to the product solution $\phi(z)\psi(r)$, yield the two equations $\phi(0) = 0$ and $\phi_z(-h) = 0$. From the first condition it follows that $B = 0$. Then the second condition yields $\cos[kh(1-a^2)^{1/2}] = 0$. The solutions of this equation are $a = a_n$ where

3.8 $$a_n = [1 - (n+\tfrac{1}{2})^2 (\tfrac{\pi}{kh})^2]^{1/2} , \quad n = 0,1,2,\cdots .$$

Thus there are infinitely many solutions of the form (3.6) satisfying (3.2) and (3.3), which we shall denote by $\phi_n(z)$, where

$$3.9 \qquad \phi_n(z) = A_n \sin[k(1-a_n^2)^{1/2}z] \ , \quad n = 0,1,2,\cdots \ .$$

Here A_n is a constant which is not yet determined.

To determine one of the constants in (3.7) we shall utilize the radiation condition. The appropriate form of this condition to select the outgoing wave is

$$3.10 \qquad \lim_{r\to\infty} r^{1/2}(\psi_r - ika\psi) = 0 \ .$$

When (3.7) is substituted into (3.10), the result is D = 0, so the outgoing solution is just a multiple of $H_o^{(1)}(kar)$. Since the product solution $\phi_n\psi$ already contains the arbitrary constant factor A_n, we can set C = 1 with no loss of generality. Then the product solution which satisfies the boundary conditions and the radiation condition is $A_n \sin[k(1-a_n^2)^{1/2}z]\ H_o^{(1)}(ka_n r)$.

Each of these product solutions is called a "normal mode", or just a "mode" for short. It is said to be propagating if a_n is real and positive, and non-propagating or evanescent if a_n is positive imaginary, because then $H_o^{(1)}(ka_n r)$ decays exponentially as r increases. From (3.8) we see that there are only M+1 propagating modes, where M is the greatest integer less than $\pi^{-1}kh - \frac{1}{2}$, and infinitely many evanescent modes.

Now to find p we represent it as a sum of modes:

$$3.11 \qquad p(r,z) = \sum_{n=0}^{\infty} A_n \sin[k(1-a_n^2)^{1/2}z]\ H_o^{(1)}(ka_n r) \ .$$

We substitute (3.11) into (3.1), using the fact that

$$3.12 \qquad \left(\frac{\partial^2}{\partial r^2} + \frac{1}{r}\frac{\partial}{\partial r} + k^2 a_n^2 \right) H_o^{(1)}(ka_n r) = \frac{4i\delta(r)}{2\pi r} \ .$$

Then (3.1) becomes

$$3.13 \qquad \sum_{n=0}^{\infty} A_n \sin[k(1-a_n^2)^{1/2}z] = \frac{i}{4}\,\delta(z-z_o) \ .$$

To solve (3.13) for A_n we multiply (3.13) by $\sin k(1-a_m^2)^{1/2}z$ and integrate it from $z = -h$ to $z = 0$. This yields

3.14 $$A_n = \frac{i}{2h} \sin[k(1-a_n^2)^{1/2} z_o] \;.$$

By using (3.14) in (3.11) we finally obtain the normal mode representation of p, which is

3.15 $$p(r,z) = \frac{i}{2h} \sum_{n=0}^{\infty} \sin[k(1-a_n^2)^{1/2} z_o] \sin[k(1-a_n^2)^{1/2} z] H_o^{(1)}(ka_n r) \;.$$

The pressure p can be conveniently calculated from (3.15), especially when kr is large. In that case the evanescent modes are negligible, and only the finite number of propagating modes need be used. We also see from (3.15) that p is symmetric in z and z_o.

3.3 Hankel transform representation

We shall now solve for p in a different way and obtain a different representation of the solution. We begin by defining the Hankel transform $\tilde{f}(s)$ of a function f(r) by

3.16 $$\tilde{f}(s) = 2\pi \int_0^{\infty} J_o(sr) f(r) r dr \;.$$

Here J_o is the Bessel function of order zero. The inverse Hankel transform is

3.17 $$f(r) = \frac{1}{2\pi} \int_0^{\infty} J_o(sr) \tilde{f}(s) s ds \;.$$

Now we multiply (3.1) - (3.3) by $2\pi J_o(kar)r$ and integrate both sides of each equation from r = 0 to r = ∞. In doing so we write $p_{rr} + r^{-1}p_r = r^{-1}(rp_r)_r$ in (3.1) and we denote the transform of p(r,z) by $\tilde{p}(s,z)$. Then we obtain

3.18 $$2\pi \int_0^{\infty} J_o(kar)(rp_r)_r dr + \tilde{p}_{zz}(ka,z) + k^2\tilde{p}(ka,z) = -\delta(z-z_o) \;,$$

3.19 $$\tilde{p}(ka,0) = 0 \;,$$

3.20 $$\tilde{p}_z(ka,-h) = 0 \;.$$

To evaluate the integral in (3.18) we require p to satisfy the radiation condition (3.10), and we choose a to have a small negative imaginary part. Then in Appendix 3.3A we show that the integral equals $-k^2a^2\tilde{p}(ka,z)$. Therefore (3.18) becomes

3.21 $$\tilde{p}_{zz} + k^2(1-a^2)\tilde{p} = -\delta(z-z_o) .$$

In order to solve (3.19) - (3.21) for $\tilde{p}(ka,z)$ we introduce two solutions of the homogeneous form of (3.21). One of them, $\tilde{p}_1$, is required to satisfy (3.19) and the other, $\tilde{p}_2$, is required to satisfy (3.20). Then we can write $\tilde{p}$ in the form

3.22 $$\tilde{p}(ka,z) = \tilde{p}_1(ka,z_>)\tilde{p}_2(ka,z_<)/W(ka) .$$

Here $z_> = \max(z,z_o)$, $z_< = \min(z,z_o)$ and $W(ka)$ is the Wronskian of $\tilde{p}_1$ and $\tilde{p}_2$. We find readily that $\tilde{p}_1 = \sin[k(1-a^2)^{1/2}z]$, $\tilde{p}_2 = \cos[k(1-a^2)^{1/2}(z+h)]$ and $W(ka) = -k(1-a^2)^{1/2}\cos[kh(1-a^2)^{1/2}]$.

Finally to obtain $p(r,z)$ we substitute the above values of $\tilde{p}_1$, $\tilde{p}_2$ and W into (3.22) for $\tilde{p}$ and then use (3.17). In this way we get

3.23 $$p(r,z) = -\frac{k}{2\pi}\int_0^\infty J_o(kar)\,\frac{\sin[k(1-a^2)^{1/2}z_>]\cos[k(1-a^2)^{1/2}(z_<+h)]}{(1-a^2)^{1/2}\cos[kh(1-a^2)^{1/2}]}\,ada .$$

This is the Hankel transform representation of p, from which p can be calculated by numerical integration.

3.3A Appendix

We shall evaluate the integral in (3.18) by defining it as the limit as $R \to \infty$ of the integral with upper limit R. Then integrating by parts twice we get

3.24 $$2\pi\int_0^R J_o(kar)(rp_r)_r dr = 2\pi J_o(kaR)Rp_r(R,z) - 2\pi ka\int_0^R J_o'(kar)rp_r dr$$

$$= 2\pi J_o(kaR)Rp_r(R,z) - 2\pi kaRJ_o'(kaR)p(R,z)$$

$$+ 2\pi ka\int_0^R [J_o'(kar)r]_r p dr .$$

When $\operatorname{Im} a < 0$, as we assume it to be, then $J_o'(kaR) \sim iJ_o(kaR)$ as $R \to \infty$.

Thus

3.25 $$2\pi J_o(kaR)Rp_r(R,z) - 2\pi kaRJ_o'(kaR)p(R,z) \sim 2\pi J_o(kaR)R[p_r(R,z)-ikap(R,z)] .$$

Because p satisfies the radiation condition (3.10), the right side of (3.25) tends to zero as $R \to \infty$. Next we use the identity $[xJ_o'(x)]' = -xJ_o(x)$ in the integral on the right side of (3.24) to write the integrand as $-kaJ_o(kar)pr$. Then the limit of the right side of (3.24) as $R \to \infty$ is just

$$-(ka)^2 \int_0^\infty J_o(kar)prdr = -(ka)^2\tilde{p}(ka,z) .$$

3.4 Ray representation

A very illuminating expression for p(r,z) is the ray representation, which we shall now obtain. To obtain it we first consider the equation (3.1) in the full three dimensional space, ignoring the boundary conditions (3.2) and (3.3). The general spherically symmetric solution of (3.1) is $p_o(R) = Ae^{ikR}/R + Be^{-ikR}/R$ where $A + B = 1/4\pi$ and $R = [r^2+(z-z_o)^2]^{1/2}$ denotes distance from the source. To eliminate the incoming wave e^{-ikR}/R we impose the radiation condition

3.26 $$\lim_{R\to\infty} R[p_o'(R)-ikp_o(R)] = 0 .$$

This condition yields B = 0 and thus the outgoing spherically symmetric solution of (3.1) in the whole space is

3.27 $$p_o(R) = e^{ikR}/4\pi R .$$

We can interpret the exponent in (3.27) as ik multiplied by the phase function R. This phase equals zero at the source and increases like the distance along a straight line from the source to the field point. We call this straight line a "ray". The factor 1/R, which multiplies the exponential factor, is called the amplitude. It decreases like the reciprocal of the square root of the cross-sectional area of a tube of rays, since that area increases like R^2. As a consequence the product of the square of the amplitude multiplied by the cross-sectional area of a ray tube remains constant along a ray. This constancy expresses the fact that energy is conserved within a ray tube. These two facts about the spherical wave (3.27) — linear increase of phase along a ray and energy conservation in a ray tube — can be used to construct the ray representation of other waves, as we shall see.

Let us now use these considerations to solve the original problem (3.1) - (3.3).

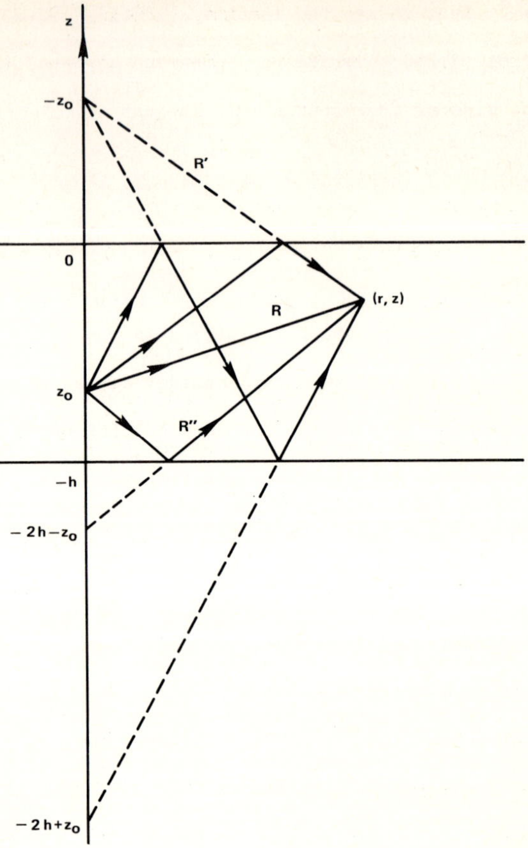

Figure 2. A point source located at $z = z_0$ emits rays in all directions. The four rays shown here all arrive at the field point (r,z). One of length R is the direct ray; another of length R' is reflected from the top surface $z = 0$ and appears to come from a source at $z = -z_0$. A third ray of length R'' is reflected from the bottom surface $z = -h$ and appears to come from a source at $z = -2h-z_0$. The fourth is reflected first from the top and then from the bottom, and appears to come from a source at $z = -2h + z_0$.

We begin with the spherical wave $p_o(R)$ given by (3.27), which satisfies (3.1) but does not satisfy the boundary conditions. When the rays associated with p_o hit the upper boundary $z = 0$, they produce reflected rays determined by the law of reflection. The phase and amplitude on each reflected ray can be found by the preceding considerations, starting with the values of the phase and amplitude on the incident ray at the point of reflection. In addition the reflected amplitude must be multiplied by a reflection coefficient equal to -1 in order that the sum of the incident and reflected waves satisfy the condition $p = 0$ on $z = 0$. Since all the reflected rays appear to come from the image source at $r = 0$, $z = -z_o$, this construction leads to a reflected wave which is just the spherical wave $-e^{ikR'}/4\pi R'$. Here R' is distance from the image source. (See Figure 2.)

A similar construction applies to the rays reflected from the bottom. However the reflection coefficient for bottom reflection is +1 because the incident and reflected waves must combine to satisfy $p_z = 0$ at $z = -h$. Furthermore the image source is at $z = -z_o - 2h$. Thus the bottom reflected wave is $e^{ikR''}/4\pi R''$ where R" is distance from the image of the source in the bottom.

Multiple reflection of the originally reflected rays gives rise to an infinite sequence of families of rays, each of which appears to come from an image point. These points are at $z = \pm z_o + 2nh$, $n = 0, \pm 1, \cdots$. By keeping track of the number of reflections from the top and bottom, we find the following expression for the total field p, which is the sum of the incident wave plus the singly and multiply reflected waves:

$$3.28 \qquad p(r,z) = \frac{1}{4\pi} \sum_{n=-\infty}^{\infty} (-1)^n \left\{ \frac{e^{ik[r^2+(z-z_o-2nh)^2]^{1/2}}}{[r^2+(z-z_o-2nh)^2]^{1/2}} - \frac{e^{ik[r^2+(z+z_o-2nh)^2]^{1/2}}}{[r^2+(z+z_o-2nh)^2]^{1/2}} \right\}.$$

This is the ray representation of the solution p.

The result (3.28) can also be derived directly by considering the successive images of the source point in the top and bottom surfaces, without considering the rays. That method of derivation, which is limited to plane boundaries and homogeneous media, is called the image method. We shall also refer to (3.28) as the <u>multiple reflection representation</u> of p, because each term in it except that with n = 0 represents a wave which has been reflected a number of times from the top and bottom boundaries.

3.5 Connections between the representations

We have now obtained three representations of the solution p(r,z) of (3.1) - (3.3). This solution determines the acoustic pressure due to a time harmonic point source in a homogeneous ocean of constant depth with a rigid bottom. The normal mode representation (3.15) is most useful at distances which are far from the source compared to the ocean depth, i.e. at distances $r >> h$. Then only the propagating modes need to be taken into account, and there are only a finite number of them. On the other hand the ray representation (3.28) is most useful near the source, where only the incident field and the first few reflected waves need to be considered because the other waves are much weaker due to spherical spreading. The Hankel transform representation (3.23) is most useful at intermediate distances. Of course all three representations are valid everywhere, but they are not equally convenient for calculation everywhere.

Since all three representations yield the same solution, they must all be equal. Therefore it must be possible to convert each representation into the other two representations. This is indeed the case, as we shall now show. The demonstration will lead to additional insight into the mathematical structure of the solution, and clarify the relation between rays and modes. Furthermore it will introduce methods of analysis which will prove useful in treating more complex problems.

Let us begin with the Hankel transform representation (3.23) in which we set $J_o = \frac{1}{2}(H_o^{(1)}+H_o^{(2)})$ to obtain

$$3.29 \quad p(r,z) = -\frac{k}{4\pi}\int_0^\infty [H_o^{(1)}(kar)+H_o^{(2)}(kar)] \frac{\sin[k(1-a^2)^{1/2}z_>]\cos[k(1-a^2)^{1/2}(z_<+h)]}{(1-a^2)^{1/2}\cos[kh(1-a^2)^{1/2}]} ada$$

Since all the functions in the integrand of (3.29) are analytic functions of a, we may interpret the integral as a line integral in the complex a plane. Therefore we can shift the path of integration to a contour C_1 from the origin to infinity slightly below the real axis out to some large real value of a, and then along the axis. (See Figure 3.) Then we use the fact that $H_o^{(2)}(kar) = -H_o^{(1)}(kae^{i\pi}r)$ to convert the integral involving $H_o^{(2)}$ along the contour C_1 to an integral involving $-H_o^{(1)}$ along $e^{i\pi}C_1$. By taking account of the minus sign multiplying $H_o^{(1)}$, and of the orientation of $e^{i\pi}C_1$, we can write (3.29) in the form

$$3.30 \quad p(r,z) = -\frac{k}{4\pi}\int_{C_1+C_2} H_o^{(1)}(kar)\frac{\sin[k(1-a^2)^{1/2}z_>]\cos[k(1-a^2)^{1/2}(z_<+h)]}{(1-a^2)^{1/2}\cos[kh(1-a^2)^{1/2}]}\,ada\ .$$

Here C_2 is $e^{i\pi}C_1$ with the orientation reversed.

We now close the contour $C_1 + C_2$ with a semi-circle Γ_R of radius R in the upper half-plane. (See Figure 3.) In the limit as $R \to \infty$, the integral (3.30) over Γ_R tends to zero, so in this limit the value of the integral is unchanged. Thus we can rewrite (3.30) as

$$3.31 \quad p(r,z) = -\frac{k}{4\pi}\lim_{R\to\infty}\int_{C_1+C_2+\Gamma} \cdots\, da\ .$$

The denominator of the integrand in (3.31) vanishes at $a = \pm 1$ and at the zeroes of $\cos[kh(1-a^2)^{1/2}]$. These zeroes are given by (3.8), and they are the poles of the integrand within the contour. The numerator vanishes at $a = \pm 1$, so these points are not poles. Then the residues of the integrand at the poles yield

$$3.32 \quad p(r,z) = -\frac{k}{4\pi}\,2\pi i\sum_{n=0}^{\infty}\frac{H_o^{(1)}(ka_nr)\sin[k(1-a_n^2)^{1/2}z_>]\cos[k(1-a_n^2)^{1/2}(z_<+h)]}{kh\sin[kh(1-a_n^2)^{1/2}]}\ .$$

To simplify (3.32) we note that $\cos[k(1-a_n^2)^{1/2}(z_<+h)] = \cos[k(1-a_n^2)^{1/2}z_<]$ $\cos[k(1-a_n^2)^{1/2}h] - \sin[k(1-a_n^2)^{1/2}z_<]\sin[k(1-a_n^2)^{1/2}h]$, and in view of (3.8), $\cos[k(1-a_n^2)^{1/2}h] = 0$. Upon using these facts in (3.32), we find that (3.32) becomes exactly (3.15), which is the normal mode representation.

This calculation provides a derivation of the normal mode representation from the Hankel transform representation. Since all the steps in the calculation are reversible, by reversing them we can derive the Hankel transform representation from the normal mode representation. These two derivations provide the connections between the Hankel transform and normal mode representations, labelled "residue" in Figure 1.

Next we shall show how to convert the ray representation (3.28) into the normal mode representation (3.15). We begin by rewriting (3.28) in the form $p(r,z) = P(r,z-z_o) - P(r,z+z_o)$ where

$$P(r,z) = \frac{1}{4\pi} \sum_{n=-\infty}^{\infty} \frac{e^{ik[r^2+(z-2nh)^2]^{1/2}-in\pi}}{[r^2+(z-2nh)^2]^{1/2}} \quad . \tag{3.33}$$

Next we rewrite the sum in (3.33) by using the Poisson summation formula [Morse and Feschbach, Methods of Theoretical Physics, p. 467, eq. (4.8.28) with $\alpha = 2\pi$]

$$\sum_{n=-\infty}^{\infty} f(2\pi n) = \frac{1}{2\pi} \sum_{q=-\infty}^{\infty} \int_{-\infty}^{\infty} f(\xi) e^{-iq\xi} d\xi \ . \tag{3.34}$$

Upon using it in (3.33) we obtain

$$P(r,z) = \frac{1}{8\pi^2} \sum_{q=-\infty}^{\infty} \int_{-\infty}^{\infty} \frac{e^{ik[r^2+(z-\xi h/\pi)^2]^{1/2}-i\xi/2-iq\xi}}{[r^2+(z-\xi h/\pi)^2]^{1/2}} d\xi \ . \tag{3.35}$$

To evaluate the integral in (3.25) we set $t = z - \xi h/\pi$ and get

$$\begin{aligned} P(r,z) &= \frac{1}{8\pi h} \sum_{q=-\infty}^{\infty} e^{-i(q+\frac{1}{2})z\pi/h} \int_{-\infty}^{\infty} e^{ik(r^2+t^2)^{1/2}+i(q+\frac{1}{2})t\pi/h} (r^2+t^2)^{-1/2} dt \\ &= \frac{i}{8h} \sum_{q=-\infty}^{\infty} e^{-i(q+\frac{1}{2})z\pi/h} H_0^{(1)}\left\{kr\left[1-([q+\tfrac{1}{2}]\pi/kh)^2\right]^{1/2}\right\} . \end{aligned} \tag{3.36}$$

Here we have used the integral representation of $H_0^{(1)}$ given in Appendix 3.5A.

We now use (3.36) for P in the relation $p(r,z) = P(r,z-z_o) - P(r,z+z_o)$. Then when we express the exponential functions of z and z_o as trigonometric functions, we obtain exactly the normal mode representation (3.15). In this way we can convert the ray representation (3.28) into the normal mode representation (3.15). Since all the steps in this calculation are reversible, the calculation also shows how the ray representation can be obtained from the normal mode representation. These calculations yield the connection labelled "Poisson summation" in Figure 1, between the ray and mode representations.

Finally we shall derive the ray representation from the Hankel transform representation (3.23). To do so we express the trigonometric functions in (3.23) in terms of exponentials and multiply them together to obtain four terms in the numerator. After dividing numerator and denominator by $\exp[ikh(1-a^2)^{1/2}]$, we can write the result in the form

$$3.37 \qquad p(r,z) = Q(r,z_>+z_<) - Q(r,-[z_>+z_<+2h]) - Q(r,z_<-z_>) + Q(r,-[z_<-z_>+2h]) \; .$$

Here $Q(r,z)$ is defined by

$$3.38 \qquad Q(r,z) = \frac{ik}{4\pi}\int_0^\infty J_o(kar)e^{ik(1-a^2)^{1/2}z}\left[1+e^{-2ikh(1-a^2)^{1/2}}\right]^{-1}(1-a^2)^{-1/2}a\,da \; .$$

We now use the binomial expansion of the factor $\left[1+e^{-2ikh(1-a^2)^{1/2}}\right]^{-1}$ in (3.38) and interchange integration and summation, which is valid, to obtain

$$3.39 \qquad Q(r,z) = \frac{ik}{4\pi}\sum_{n=0}^\infty (-1)^n\int_0^\infty J_o(kar)e^{ik(1-a^2)^{1/2}(z-2nh)}(1-a^2)^{-1/2}a\,da \; .$$

The integral in (3.39) is evaluated in Appendix 3.5A. When the result (3.45) is used in (3.39) it yields

$$3.40 \qquad Q(r,z) = \frac{-1}{4\pi}\sum_{n=0}^\infty (-1)^n[r^2+(z-2nh)^2]^{-1/2}\, e^{ik[r^2+(z-2nh)^2]^{1/2}} \; .$$

Now we substitute (3.40) into (3.37) and note that the first two terms in (3.37) combine to yield the last term in (3.28), while the second two terms in (3.37) combine to yield the first term in (3.28). Thus (3.28) follows from (3.37).

This calculation shows how the ray representation (3.28) can be obtained from the Hankel transform representation (3.23). Since all the steps are reversible, it also shows how the Hankel transform representation follows from the ray representation. Thus this calculation yields the connection between these two representations, labelled "binomial expansion" in Figure 1. This completes the demonstration of all the connections between the representations indicated in Figure 1. Most of the derivations indicated in Figure 1 are given by Brekhovskikh [1] Chapter V, who also discusses various properties of the solution.

3.5A Appendix

The integral in (3.36) can be evaluated by first setting $t = r \sinh\theta$ and $m = (q + \frac{1}{2})\pi/kh$ to obtain

$$3.41 \quad \int_{-\infty}^{\infty} e^{ik[(r^2+t^2)^{1/2}+(q + \frac{1}{2})\pi t/kh]}(r^2+t^2)^{-1/2}dt = \int_{-\infty}^{\infty} e^{ikr[\cosh\theta+m \sinh\theta]}d\theta .$$

Now $\cosh\theta + m \sinh\theta = (1-m^2)^{1/2}\cosh[\theta+\tanh^{-1}m] = (1-m^2)^{1/2}\cosh\theta'$ where $\theta' = \theta + \tanh^{-1}m$. We next rewrite the last integral, simplifying the exponent with the aid of this relation, and then we set $\sinh\theta' = s/r$ to get

$$3.42 \quad \int_{-\infty}^{\infty} e^{ikr(1-m^2)^{1/2}\cosh\theta'}d\theta' = \int_{-\infty}^{\infty} e^{ik(1-m^2)^{1/2}(r^2+s^2)^{1/2}}(r^2+s^2)^{-1/2}ds .$$

According to Magnus and Oberhettinger [2] page 27, the last integral is just $i\pi H_0^{(1)}[kr(1-m^2)^{1/2}]$, and thus this is the value of the first integral in (3.41).

To calculate the integral in (3.39) we begin with equations 5 and 6 on page 761 of Gradshteyn and Ryzhik [3] . By adding i times equation 5 to equation 6 we get

3.43
$$\int_0^\infty x(x^2+z^2)^{-1/2}e^{ik(x^2+z^2)^{1/2}}J_o(kax)dx = (\pi z/2k)^{1/2}(1-a^2)^{-1/4}$$

$$\left\{-N_{-1/2}(kz[1-a^2]^{1/2}) + iJ_{-1/2}(kz[1-a^2]^{1/2})\right\}, \quad 0 < a < 1$$

$$= (2z/\pi k)^{1/2}(a^2-1)^{-1/4}K_{1/2}(kz[a^2-1]^{1/2}) \ , \quad 1 < a \ .$$

Next we use the expressions for $J_{-1/2}$, $N_{-1/2}$ and $K_{1/2}$ given on pages 437, 438 and 443 of Abramowitz and Stegun [4] , in (3.43) to obtain

3.44
$$\int_0^\infty x(x^2+z^2)^{-1/2}e^{ik(x^2+z^2)^{1/2}}J_o(kax)dx = \frac{-i}{k}(1-a^2)^{-1/2}e^{ikz(1-a^2)^{1/2}} \ .$$

Since (3.44) is of the form (3.16), it is a Hankel transform. Then the inverse Hankel transform (3.17) yields

3.45
$$\int_0^\infty (1-a^2)^{-1/2}e^{ikz(1-a^2)^{1/2}}J_o(kax)ada = \frac{i}{k}(x^2+z^2)^{-1/2}e^{ik(x^2+z^2)^{1/2}} \ .$$

The left side of (3.45) is the integral in (3.39), so (3.45) is the desired evaluation of it.

4. The inhomogeneous stratified ocean of constant depth

4.1 Introduction

We shall now consider the acoustic pressure produced by a time harmonic point source in a stratified inhomogeneous ocean of constant depth. If the source location $\underline{x}_o$ is taken to be $r = 0$, $z = z_o$ in cylindrical coordinates, then (2.3) becomes

4.1 $$\Delta p + k^2 n^2(z) p = -\delta(z-z_o)\delta(r)/2\pi r \ .$$

The surface condition is (2.4),

4.2a $$p = 0 \quad \text{at} \quad z = 0 \ .$$

Since the depth h is constant, the bottom condition (2.5) becomes

4.2b $$p_z = 0 \quad \text{at} \quad z = -h \ .$$

We seek that solution of (4.1) - (4.3) which satisfies a suitable radiation condition, which we shall state explicitly later. Since the problem is axially symmetric, the solution $p(r,z)$ is independent of the angular coordinate θ .

Just as in section 3, we shall obtain three representations of the solution. The normal mode and Hankel transform representations are similar to the corresponding ones of section 3, while the multiple scattering representation corresponds to the previous ray representation. This representation involves the effects of upward and downward refraction due to inhomogeneity, as well as the effects of reflection from the top and bottom. We use the term "scattering" in describing it to indicate that it includes both of these effects. The ray representation of section 3 involved only reflection, because the ocean was assumed to be homogeneous. The three representations are indicated in Figure 4.

After obtaining the three representations, we shall show how they can be transformed into one another. These transformations are also indicated in Figure 4. Then in section 5 we shall obtain asymptotic expansions of these representations. They

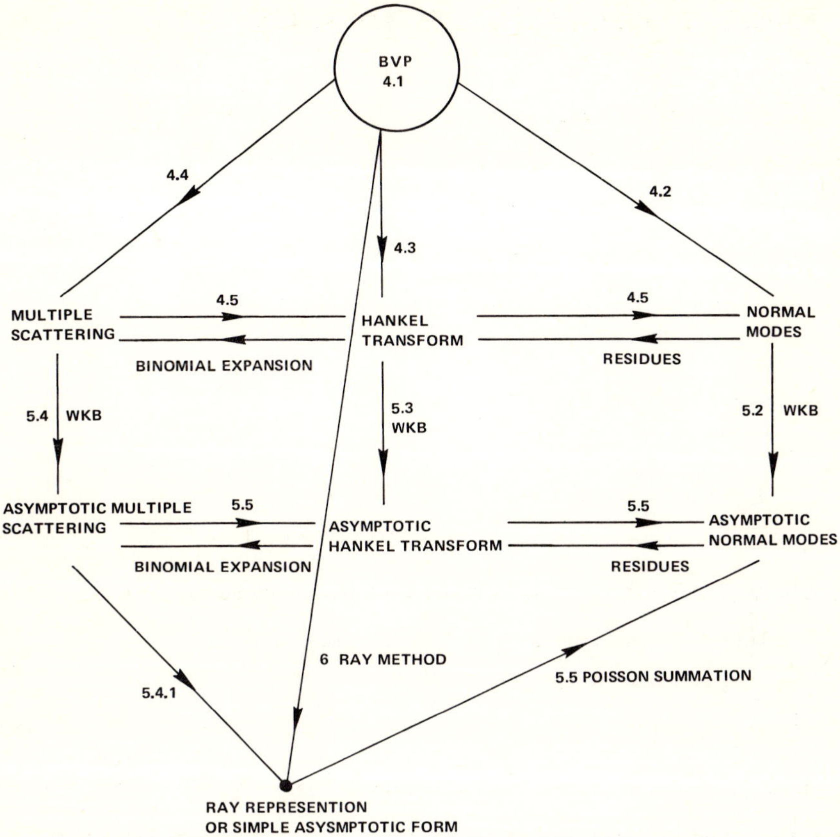

Figure 4. The boundary value problem is solved by the method of normal modes in subsection 4.2, by Hankel transformation in 4.3, and by the multiple scattering method in 4.4. This yields the three representations indicated by normal modes, Hankel transform and multiple scattering, respectively. These representations are transformed into one another in 4.5. Then with the aid of the WKB method, asymptotic forms of these representations are obtained in 5.2, 5.3 and 5.4. They are converted into one another in subsection 5.5. A simpler asymptotic form of the multiple scattering representation is deduced in 5.4.1. This same representation is derived directly by the ray method in section 6.

lead to a better understanding of the solution and are also useful for calculation.

4.2 Normal mode representation

As in sub-section 3.2, we shall seek the solution $p(r,z)$ of (4.1) and (4.2a,b) as a sum of normal modes. Each normal mode is a product solution $\phi(z)\psi(r)$ of the homogeneous equations. Substitution of such a product into the homogeneous form of (4.1) and separation of variables yields the two equations

4.3 $$\phi_{zz} + k^2n^2(z)\phi = k^2a^2\phi ,$$

4.4 $$\psi_{rr} + r^{-1}\psi_r = -k^2a^2\psi .$$

The separation constant has been written as ka for convenience. When the product solution is substituted into the boundary conditions (4.2a,b), they become

4.5 $$\phi(0) = 0 , \quad \phi_z(-h) = 0 .$$

The problem (4.3) and (4.5) has an infinite number of simple real eigenvalues $a_0^2 > a_1^2 > a_2^2 > \cdots$ and corresponding eigenfunctions $\phi_0, \phi_1, \cdots$ which form a complete orthogonal set. There are a finite number of positive eigenvalues and infinitely many negative ones, with a_n^2 tending to $-\infty$ as n increases. We shall take $a_n > 0$ if $a_n^2 > 0$ and $\mathrm{Im}\, a_n > 0$ if $a_n^2 < 0$. Then the solution $\psi(r)$ of (4.4), which satisfies the radiation condition (3.10) with $a = a_n$, is a constant multiple of $H_0^{(1)}(ka_n r)$, just as in sub-section 3.2. Thus the normal modes are $A_n\phi_n(z)H_0^{(1)}(ka_n r)$, where A_n is an arbitrary constant.

In view of the completeness of the $\phi_n(z)$, we can write $p(r,z)$ in the form

4.6 $$p(r,z) = \sum_{n=0}^{\infty} A_n\phi_n(z)H_0^{(1)}(ka_n r) .$$

To find the A_n we substitute (4.6) into (4.1) and use the relation (3.12) to obtain

$$4.7 \qquad \sum_{n=0}^{\infty} A_n\phi_n(z) = \frac{i}{4}\,\delta(z-z_o)\ .$$

We now multiply (4.7) by $\phi_m(z)$ and then integrate the result from $z = -h$ to $z = 0$. Because the $\phi_n(z)$ are orthogonal, this yields

$$4.8 \qquad A_n = \frac{i}{4}\,\phi_n(z_o) \Big/ \int_{-h}^{0} \phi_n^2(z)dz\ .$$

Now (4.6) can be written as

$$4.9 \qquad p(r,z) = \frac{i}{4} \sum_{n=0}^{\infty} \phi_n(z)\phi_n(z_o)H_o^{(1)}(ka_n r) \Big/ \int_{-h}^{0} \phi_n^2(s)ds\ .$$

This is the normal mode representation of $p(r,z)$.

Just as in sub-section 3.2, the finite number of terms in (4.9) with a_n real represent propagating modes, while the remaining terms represent evanescent or non-propagating modes. For kr large, only the propagating modes are significant, so then p can be calculated easily from (4.9). For all r, (4.9) shows that p is symmetric in z and z_o .

4.3 Hankel transform representation

Now we shall proceed as in sub-section 3.3 to obtain the Hankel transform representation of the outgoing solution of (4.1) and (4.2a,b). First we multiply each equation by $2\pi J_o(kar)$ and then integrate the result from $r = 0$ to $r = \infty$. We also denote the Hankel transform of $p(r,z)$ by $\tilde{p}(s,z)$. In integrating (4.1) we use the result of Appendix 3.3A. In this way we obtain from (4.1) - (4.3) the transformed equations

$$4.10 \qquad \tilde{p}_{zz} + k^2[n^2(z)-a^2]\tilde{p} = -\delta(z-z_o)\ ,$$

$$4.11 \qquad \tilde{p}(ka,0) = 0\ ,$$

$$4.12 \qquad \tilde{p}_z(ka,-h) = 0\ .$$

We now introduce two solutions $\tilde{p}_1(ka,z)$ and $\tilde{p}_2(ka,z)$ of the homogeneous form of (4.10), which satisfy (4.11) and (4.12) respectively. Then the solution of (4.10) - (4.12) is readily found to be of the same form as (3.22), i.e.

4.13 $$\tilde{p}(ka,z) = \tilde{p}_1(ka,z_>)\tilde{p}_2(ka,z_<) \Big/ W(ka) .$$

By applying the inverse Hankel transform (3.17) to (4.13) we get

4.14 $$p(r,z) = \frac{k^2}{2\pi} \int_0^\infty J_o(kar)\tilde{p}_1(ka,z_>)\tilde{p}_2(ka,z_<)[W(ka)]^{-1} a\,da .$$

This is the Hankel transform representation of p(r,z), which can be used to calculate p numerically.

4.4 Multiple scattering representation

A third representation of p can be obtained by considering the propagation of waves outward from the source, and taking account successively of their refraction by the medium and reflection by the boundaries. Since both reflection and refraction are special kinds of scattering, and since they occur repeatedly, we call this the multiple scattering representation. We shall write it in the form

4.15 $$p(r,z) = \sum_{n=0}^{\infty} q_n(r,z) .$$

Here q_o represents the direct wave from the source, q_1 represents a wave which has been scattered (i.e. reflected or refracted) once, and q_n represents a wave which has been scattered n times. The series (3.28) for p in a homogeneous ocean is exactly of the form (4.15) if the terms with ±n are combined to yield q_n . In that case only reflection occurs, and the multiple scattering representation becomes just the multiple reflection representation.

One way to obtain a multiple scattering representation of p in an inhomogeneous medium is to express p as a Born expansion in powers of $n^2(z) - n^2(z_o)$. Then the term q_n is an n-fold integral over the ocean, and therefore it is not easy to evaluate.

Another procedure, which we shall use, is to obtain a representation of the form (4.15) in which q_n is only asymptotically an n times scattered wave. Here asymptotic refers to the limit of small wavelength $\lambda = 2\pi/k$, i.e. to k large. To find this representation we first apply the Hankel transform to the problem, thus converting it to (4.10) - (4.13). Then we seek a multiple scattering representation of the solution $\tilde{p}$ of this problem.

In order to construct this representation, we introduce two particular solutions of the homogeneous form of (4.10), which we denote U(ka,z) and D(ka,z) . The solutions U(ka,z) and D(ka,z) are characterized by the properties that asymptotically for ka large, they represent upward and downward traveling waves in the neighborhood of z_o, respectively. We normalize them so that their Wronskian W(U,D) has the value

4.16 $$W(U,D) = -2ik \ .$$

We now use U and D to construct the term q_o in the Hankel transform of the series (4.15), by solving (4.10) without regard to the boundary conditions. We readily find

4.17 $$\tilde{q}_o(ka,z) = U(ka,z_>)D(ka,z_<)/(-2ik) \ .$$

When $\tilde{q}_o$ is incident upon the upper boundary z = 0 , it produces a downward traveling wave proportional to D(ka,z), which we shall write as $R_1(ka)D(ka,z_o)D(ka,z)/(-2ik)$. To find R_1 we set $\tilde{q}_o + R_1 D(ka,z_o)D(ka,z)/(-2ik) = 0$ at z = 0 in accordance with (4.11), and find that R_1 is given by

4.18 $$R_1(ka) = -U(ka,0)/D(ka,0) \ .$$

Similarly, when $\tilde{q}_o$ is incident upon the lower boundary z = -h it produces an upward traveling wave which we shall write as $R_2(ka)U(ka,z_o)U(ka,z)/(-2ik)$. Then we set $\partial_z\tilde{q}_o + R_2 U(ka,z_o)\partial_z U(ka,z)/(-2ik) = 0$ at z = -h to satisfy (4.12). This yields

$$4.19 \qquad R_2(ka,z) = -\partial_z D(ka,-h)/\partial_z U(ka,-h) .$$

Upon adding together the two reflected waves produced by $\tilde{q}_o$ we get $\tilde{q}_1$, which is given by

$$4.20 \qquad \tilde{q}_1(ka,z) = \frac{-1}{2ik} [R_1(ka)D(ka,z_o)D(ka,z)+R_2(ka)U(ka,z_o)U(ka,z)] .$$

By continuing to calculate the successively scattered waves in the same way, we find that $\tilde{q}_n$ is given by the following formulas, from which the argument ka is omitted

$$4.21 \qquad \tilde{q}_{2m}(ka,z) = \frac{-1}{2ik} [R_1^m R_2^m U(z_>)D(z_<)+R_1^{m+1}R_2^{m+1}U(z_<)D(z_>)] ,$$

$$\tilde{q}_{2m+1}(ka,z) = \frac{-1}{2ik} [R_1^{m+1}R_2^m D(z_>)D(z_<)+R_1^m R_2^{m+1}U(z_<)U(z_>)] .$$

We now sum the $\tilde{q}_n$ given by (4.21) to obtain $\tilde{p}(ka,z)$. Then we apply the inverse Hankel transform (3.17) to the sum to obtain $p(r,z)$. Upon interchanging the order of summation and integration, we find

$$4.22 \qquad p(r,z) = \sum_{m=0}^{\infty} \frac{ik}{4\pi} \int_0^{\infty} J_o(kar) \Big[(R_1R_2)^m U(ka,z_>)D(ka,z_<) + (R_1R_2)^{m+1}U(ka,z_<)D(ka,z_>) + R_1^{m+1}R_2^m D(ka,z_>)D(ka,z_<) + R_1^m R_2^{m+1}U(ka,z_<)U(ka,z_>) \Big] ada .$$

In (4.22) R_1 and R_2 are given by (4.18) and (4.19).

The result (4.22) is the multiple scattering representation of $p(r,z)$. The four types of terms in the integrand have the following interpretations for $z > z_o$: The term $(R_1R_2)^m U(z)D(z_o)$ represents a wave which travels upward from the source, is reflected m times at each boundary, and is traveling upward at z. The term $(R_1R_2)^{m+1}U(z_o)D(z)$ represents a wave which travels downward from the source, is reflected m+1 times at each boundary, and is traveling downward at z. The other two terms represent waves which leave the source going in one direction and pass through z in the opposite direction after m reflections from one boundary and m+1 reflections from the other boundary. Wherever we have said a wave is reflected, we include the possibility that instead it is turned by refraction before reaching the boundary. This will become evident when we determine the asymptotic form of each term in the next section.

4.5 Connections between the representations

We have obtained three representations of the solution $p(r,z)$ of (4.1) - (4.3). Now we shall show how they can be transformed into one another. We shall first show how the Hankel transform representation (4.14) can be transformed into the multiple scattering representation (4.22). To do so we note that the functions $\tilde{p}_1$ and $\tilde{p}_2$ in (4.14) can be expressed as follows in terms of the functions U, D, R_1 and R_2 which occur in (4.22):

$$\tilde{p}_1(ka,z) = U(ka,z) + R_1(ka)D(ka,z) \;, \tag{4.23}$$

$$\tilde{p}_2(ka,z) = D(ka,z) + R_2(ka)U(ka,z) \;. \tag{4.24}$$

To verify these relations we note that both sides of each equation are solutions of the homogeneous form of (4.10). Furthermore from the definition (4.18) of R_1 it follows that the right side of (4.23) satisfies (4.11), while from the definition (4.19) of R_2 the right side of (4.24) satisfies (4.12). Thus $\tilde{p}_1$ and $\tilde{p}_2$ can be defined by (4.23) and (4.24).

We next substitute these expressions for $\tilde{p}_1$ and $\tilde{p}_2$ into (4.14). In doing so we note that $W(\tilde{p}_1,\tilde{p}_2) = (1-R_1R_2)W(U,D) = -2ik(1-R_1R_2)$. Then (4.14) becomes

$$p(r,z) = \frac{ik}{4\pi}\int_0^\infty J_o(kar)[U(ka,z_>)+R_1D(ka,z_>)][D(ka,z_<)+R_2U(ka,z_<)] \\ (1-R_1R_2)^{-1}\, ada \;. \tag{4.25}$$

Upon expanding $(1-R_1R_2)^{-1}$ by the binomial theorem, and then interchanging the order of summation and integration, we find that (4.25) becomes exactly (4.22). Thus the Hankel transform representation has been transformed into the multiple scattering representation. Since all the steps in the transformation are reversible, the reversal of them yields the former representation from the latter. These transformations are indicated in Figure 4, where they are labelled "binomial expansion".

Now we shall show how the Hankel transform representation (4.14) can be converted into the normal mode representation (4.9). First by proceeding as in section 3.5 we rewrite (4.14) in the following form, which is analogous to (3.31):

$$4.26 \qquad p(r,z) = \frac{k^2}{4\pi} \lim_{R\to\infty} \int_{C_1+C_2+\Gamma} H_o^{(1)}(kar)\tilde{p}_1(ka,z_>)\tilde{p}_2(ka,z_<)W^{-1}(ka)ada \ .$$

The contour $C_1 + C_2 + \Gamma$ is shown in Figure 3. Next we denote by a_m , $m = 0,1,2,\ldots$ the roots of the equation $W(ka) = 0$ which lie in the upper half of the a-plane. These roots are all simple and they are the only poles of the integrand of the integral in (4.26) in the upper half-plane. Therefore a residue evaluation of that integral yields

$$4.27 \qquad p(r,z) = \frac{ik}{2} \sum_{n=0}^{\infty} H_o^{(1)}(ka_nr)\tilde{p}_1(ka_n,z_>)\tilde{p}_2(ka_n,z_<)/W'(ka_n) \ .$$

Since $W(ka_n) = 0$, it follows that $\tilde{p}_1(ka_n,z)$ is a multiple of $\tilde{p}_2(ka_n,z)$. Therefore both $\tilde{p}_1$ and $\tilde{p}_2$ are multiples of the eigenfunction $\phi_n(z)$, so we shall write $\tilde{p}_1(ka_n,z) = \phi_n(z)$ and $\tilde{p}_2(ka_n,z) = \alpha_n\phi_n(z)$ where α_n is a constant. In Appendix 4.5A we show that

$$4.28 \qquad W'(ka_n) = 2k\alpha_n a_n \int_{-h}^{0} \phi_n^2(z)dz \ .$$

When these results are used in (4.27) it becomes (4.9). Thus the normal mode expansion is obtained from the Hankel transform representation. By reversing the steps, the latter representation can be obtained from the former. These transformations are indicated by the line labelled "residues" in Figure 4. This completes the derivation of the relations shown in that figure.

4.5A Appendix

We shall now evaluate $W'(ka_n)$ where $W(ka) = \tilde{p}_1 \partial_z \tilde{p}_2 - \tilde{p}_2 \partial_z \tilde{p}_1$. Since W is independent of z, we can evaluate it at $z = 0$ where $\tilde{p}_1 = 0$ to get $W(ka) = -\tilde{p}_2(ka,0)\partial_z \tilde{p}_1(ka,0)$. We next write the homogeneous form of (4.10), which is satisfied by $\tilde{p}_2(ka,z)$, and write (4.3) with $a = a_n$, which is satisfied by $\phi_n(z)$. We multiply the first equation by $\phi_n(z)$ and the second by $\tilde{p}_2(ka,z)$ and then substract the two to obtain

$$\partial_z[\phi_n(z)\partial_z \tilde{p}_2(ka,z) - \tilde{p}_2(ka,z)\partial_z \phi_n(z)] = k^2(a^2-a_n^2)\tilde{p}_2(ka,z)\phi_n(z) \ . \tag{4.29}$$

Upon integrating (4.29) with respect to z from $z = -h$ to $z = 0$, and using the boundary conditions (4.5) and (4.12), we get

$$-\tilde{p}_2(ka,0)\partial_z \phi_n(0) = k^2(a^2-a_n^2) \int_{-h}^{0} \tilde{p}_2(ka,z)\phi_n(z)dz \ . \tag{4.30}$$

Now we solve for $\tilde{p}_2(ka,0)$ in the expression above for $W(ka)$ and substitute the result into (4.30). Then by rearranging factors, we get

$$\frac{W(ka)}{k(a-a_n)} = k(a+a_n) \frac{\partial_z \tilde{p}_1(ka,0)}{\partial_z \phi_n(0)} \int_{-h}^{0} \tilde{p}_2(ka,z)\phi_n(z)dz \ . \tag{4.31}$$

Finally we let a tend to a_n and the left side of (4.31) becomes $W'(ka)$, since $W(ka_n) = 0$, while the right side becomes the right side of (4.28). This proves (4.28).

5. Asymptotic representations for an inhomogeneous stratified ocean of constant depth

5.1 Introduction

We have obtained three representations of the acoustic pressure due to a point source in an inhomogeneous stratified ocean of constant depth. Now we shall obtain the asymptotic forms of those representations, which are valid when the wavelength is small compared to the other lengths of the problem. These other lengths are the range r, the bottom depth h, the source depth z_o, and the vertical distance over which the sound velocity changes appreciably. Analytically this is equivalent to assuming that k is large. The asymptotic forms are simpler to use, easier to calculate with, and permit an intuitively appealing interpretation of the results.

The asymptotic forms of the modal and Hankel transform representations involve the WKB asymptotic forms of the solutions of certain ordinary differential equations. The asymptotic form of the multiple scattering representation involves the stationary phase evaluation of certain integrals. The result has a direct interpretation in terms of the rays of geometrical acoustics. After deriving these asymptotic forms, we shall show the connections between them.

In section 6 we give a direct geometrical derivation of the asymptotic representation by means of rays. This ray representation will be shown to be the same as the asymptotic form of the multiple scattering representation, which thus provides a justification of it. The main virtue of the ray method of derivation is that it also applies to nonstratified inhomogeneous oceans of variable depth.

As we shall see, the asymptotic results depend upon the form of the function n(z) and the source depth z_o . We shall assume that n(z) has the form shown in Figure 5, and that the source is located in the sound channel, as is indicated in the figure.

5.2 Asymptotic form of the modal representation

The representation of p(r,z) in terms of normal modes is given by (4.9) which is

5.1 $$p(r,z) = \frac{i}{4}\sum_{m=0}^{\infty} \phi_m(z)\phi_m(z_o)H_o^{(1)}(ka_m r) \Big/ \int_{-h}^{0} \phi_m^2(s)ds \ .$$

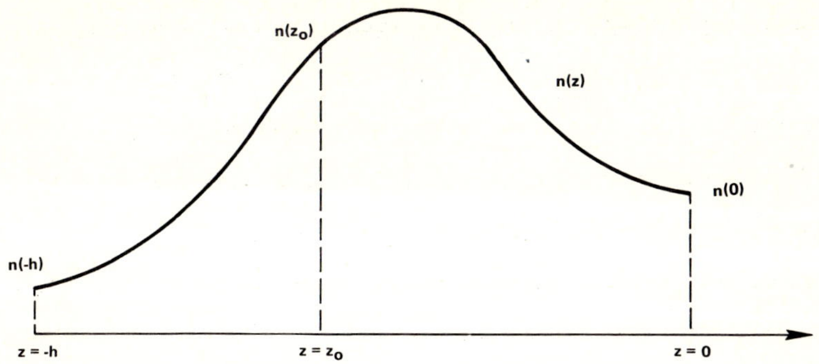

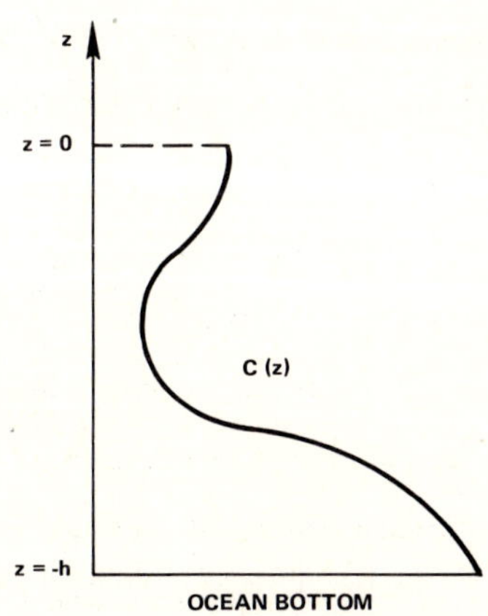

Figure 5. The qualitative form of the refractive index profile n(z) assumed in section 5. The source is supposed to lie within the sound channel.

The eigenvalues a_m and eigenfunctions $\phi_m(z)$ are solutions of (4.3) and (4.5). Their asymptotic forms for k large are determined in Appendix 5.2A by the WKB method. There are three different asymptotic forms, each applicable to one of the intervals $[0,n(-h))$, $(n(-h),n(0))$, $(n(0),n_{max})]$ within which the real eigenvalues lie. Here $n_{max} = \max n(z)$. The values of the integer m for which each form applies are those for which the corresponding eigenvalue lies in the appropriate interval.

The three asymptotic forms are listed below, together with the eigenvalue equation and the values of the normalization integrals obtained by using them:

$\underline{0 \le a_m < n(-h)}$

5.2 $$\phi_m(z) \sim (n^2-a_m^2)^{-1/4} \cos\left[k \int_{-h}^{z} (n^2-a_m^2)^{1/2}dz'\right] , \quad -h \le z \le 0 ,$$

5.3 $$k \int_{-h}^{0} (n^2-a_m^2)^{1/2}dz = (m + \tfrac{1}{2})\pi ,$$

5.4 $$\int_{-h}^{0} \phi_m^2(z)dz \sim \frac{1}{2} \int_{-h}^{0} (n^2-a_m^2)^{-1/2}dz .$$

$\underline{n(-h) < a_m < n(0)}$

5.5 $$\phi_m(z) \sim (n^2-a_m^2)^{-1/4} \cos\left[k \int_{z_m}^{z} (n^2-a_m^2)^{1/2}dz' - \frac{\pi}{4}\right] , \quad z_m < z \le 0 ,$$

$$\sim 0 \qquad , \quad -h \le z < z_m ,$$

5.6 $$k \int_{z_m}^{0} (n^2-a_m^2)^{1/2}dz = (m + \tfrac{3}{4})\pi ,$$

5.7 $$\int_{-h}^{0} \phi_m^2 dz \sim \frac{1}{2} \int_{z_m}^{0} (n^2-a_m^2)^{-1/2} dz \quad .$$

Here z_m is the root of $n(z_m) = a_m$.

<u>$n(0) < a_m < n_{max}$</u>

5.8 $$\phi_m(z) \sim (n^2-a_m^2)^{-1/4} \cos\left[k \int_{z_m}^{z} (n^2-a_m^2)^{1/2} dz' - \frac{\pi}{4}\right], \quad z_m < z < z'_m \ ,$$

$$\sim 0 \quad , \quad -h \leq z < z_m \ ,$$

$$\sim 0 \quad , \quad z'_m < z \leq 0 \ ,$$

5.9 $$k \int_{z_m}^{z'_m} (n^2-a_m^2)^{1/2} dz = (m + \tfrac{1}{2})\pi \quad ,$$

5.10 $$\int_{-h}^{0} \phi_m^2 dz \sim \frac{1}{2} \int_{z_m}^{z'_m} (n^2-a_m^2)^{-1/2} dz \quad .$$

Here z_m is the smaller and z'_m the larger root of $n(z) = a_m$. The above formulas are not valid at the turning points z_m and z'_m . In addition, we shall use the following asymptotic form of the Hankel function:

5.11 $$H_o^{(1)}(ka_m r) \sim (2/\pi k a_m r)^{1/2} \, e^{i(ka_m r-\pi/4)} \quad .$$

We now substitute the above formulas into (5.1). The result can be written in the following form:

5.12 $$p(r,z) \sim \frac{1}{(2\pi kr)^{1/2}}$$

$$\times \sum_{\substack{a_m > n(-h) \\ z_m < z, z_o < z'_m}} \frac{e^{ika_m r + i\pi/4} \cos\left[k \int_{z_m}^{z} (n^2-a_m^2)^{1/2} dz - \pi/4\right] \cos\left[k \int_{z_m}^{z_o} (n^2-a_m^2)^{1/2} dz - \pi/4\right]}{[n^2(z)-a_m^2]^{1/4} [n^2(z_o)-a_m^2]^{1/4} \int_{z_m}^{z'_m} [a_m/(n^2-a_m^2)]^{1/2} dz}$$

$$+ \frac{1}{(2\pi kr)^{1/2}} \sum_{a_m < n(-h)} \frac{e^{ika_m r + i\pi/4} \cos\left[k \int_{-h}^{z} (n^2-a_m^2)^{1/2} dz\right] \cos\left[k \int_{-h}^{z_o} (n^2-a_m^2)^{1/2} dz\right]}{[n^2(z)-a_m^2]^{1/4} [n^2(z_o)-a_m^2]^{1/4} \int_{-h}^{0} [a_m/(n^2-a_m^2)]^{1/2} dz}$$

Here we define $z'_m = 0+$ when there is only one root of $n(z) = a_m$. This result (5.12) is the asymptotic form of the normal mode representation of $p(r,z)$. Only a finite number of modes are propagating, and for kr large they are the only ones that need be taken into account in evaluating the sum.

The result (5.12) is not valid when either z or z_o is equal to either z_m or z'_m , because the asymptotic forms of the $\phi_m(z)$ given above are not valid then. This defect can be overcome by using other asymptotic forms of $\phi_m(z)$ in thin boundary layers around the turning points. These boundary layer asymptotic forms, which involve Airy functions, are used to derive the WKB connection formulas employed in Appendix 5.2A in getting the above asymptotic forms of $\phi_m(z)$. A different method of overcoming the defects, which we shall use instead, is to represent $\phi_m(z)$ by a uniform asymptotic form valid both at and near a turning point as well as away from it. Such a uniform asymptotic form is derived in Appendix 5.3B, and also involves Airy functions. When there are two turning points, this asymptotic form is only semi-uniform, since two different forms are valid around the two turning points.

The uniform or semi-uniform asymptotic forms of $\phi_m(z)$ just described, together with the corresponding equations for the eigenvalues a_m , are:

$\underline{0 \le a_m < n(0)}$

5.13 $$\phi_m(z) \sim [S_z(z)]^{-1/2}\left\{Ai[-k^{2/3}S(z)]-Ai'[-k^{2/3}S(-h)]Bi[-k^{2/3}S(z)]/Bi'[-k^{2/3}S(-h)]\right\}$$

5.14 $$Ai[-k^{2/3}S(0)]Bi'[-k^{2/3}S(-h)] - Bi[-k^{2/3}S(0)]Ai'[-k^{2/3}S(-h)] = 0 \ .$$

Here S(z) is defined by

5.15 $$S(z) = \left\{\frac{3}{2}\int_{z_m}^{z} [n^2(z')-a_m^2]^{1/2}dz'\right\}^{2/3} \ .$$

In the interval $0 \le a_m < n(-h)$, (5.13) and (5.14) are asymptotically equivalent to the simpler WKB formulas (5.2) and (5.3), but the latter are not uniformly valid for a_m close to n(-h) . When $-k^{-2/3}S(-h) >> 1$, (5.13) and (5.14) simplify to

5.16 $$\phi_m(z) \sim [S_z(z)]^{-1/2}\ Ai[-k^{2/3}S(z)] \ ,$$

5.17 $$Ai[-k^{2/3}S(0)] = 0 \ .$$

In the interval $0 \le a_m < n(-h)$, the turning point z_m lies to the left of -h (i.e. $z_m < -h$) , and is obtained by continuing n(z) into the interval z < -h .

$\underline{n(0) \le a_m < n_{max}}$

For $-h \le z < z_m'$, (5.13) holds with a_m determined by (5.21). For $z_m < z \le 0$, $\phi_m(z)$ is given by

$$5.18 \quad \phi_m(z) \sim c_m[-\bar{S}_z(z)]^{-1/2}\left\{Ai[-k^{2/3}\bar{S}(z)]-Ai[-k^{2/3}\bar{S}(0)]Bi[-k^{2/3}\bar{S}(z)]/Bi[-k^{2/3}\bar{S}(0)]\right\}.$$

Here $\bar{S}(z)$ and c_m are defined by

$$5.19 \qquad \bar{S}(z) = \left\{\frac{3}{2}\int_z^{z'_m}[n^2(z')-a_m]^{1/2}dz'\right\}^{2/3},$$

$$5.20 \quad c_m=\left[\sin\left(k\int_{z_m}^{z'_m}[n^2(z')-a_m^2]^{1/2}dz'\right)-\frac{Ai[-k^{2/3}\bar{S}(0)]}{Bi[-k^{2/3}\bar{S}(0)]}\cos\left(k\int_{z_m}^{z'_m}[n^2(z')-a_m^2]^{1/2}dz'\right)\right]^{-1}.$$

The equation for a_m is

$$5.21 \quad \cot\left\{k\int_{z_m}^{z'_m}[n^2(z')-a_m^2]^{1/2}dz'\right\}=\frac{Bi[-k^{2/3}\bar{S}(0)]Ai'[-k^{2/3}S(-h)]-Ai[-k^{2/3}\bar{S}(0)]Bi'[-k^{2/3}S(-h)]}{Ai[-k^{2/3}\bar{S}(0)]Ai'[-k^{2/3}S(-h)]-Bi[-k^{2/3}\bar{S}(0)]Bi'[-k^{2/3}S(-h)]}.$$

When $-k^{-2/3}\bar{S}(0) >> 1$ and $-k^{-2/3}S(-h) >> 1$, then $c_m \sim (-1)^m$, (5.21) reduces to (5.9), and

$$5.22 \qquad \phi_m(z) \sim (-1)^m\,[-\bar{S}_z(z)]^{-1/2}\,Ai[-k^{2/3}\bar{S}(z)].$$

The two asymptotic forms (5.13) and (5.18) are asymptotically equal in the inverval $z_m < z < z'_m$, where they are both valid. Neither is valid for a_m near n_{max}, when the two turning points are close together.

To obtain the semi-uniform asymptotic form of the modal representation of $p(r,z)$, we use (5.13) and (5.18) for ϕ_m in (5.1), together with (5.11) for $H_o^{(1)}$. When both z and z_o are below the height z_{max} at which $n(z)$ attains its maximum, ϕ_m is given by (5.13). If in addition $-k^{-2/3}S(-h) >> 1$, and $-k^{-2/3}\bar{S}(0) >> 1$, then (5.1) becomes

5.23
$$p(r,z) \sim \frac{1}{(8\pi kr)^{1/2}} \times \sum_m e^{ika_m r + i\pi/4} \frac{[a_m S_z(z) S_z(z_o)]^{-1/2} Ai[-k^{2/3}S(z)] Ai[-k^{2/3}S(z_o)]}{\int_{-h}^{z_{max}} [S_z(z)]^{-1} Ai^2[-k^{2/3}S(z)]dz - \int_{z_{max}}^{0} [\bar{S}_z(z)]^{-1} Ai^2[-k^{2/3}\bar{S}(z)]dz}$$

Similar but more complicated formulas hold for other ranges of z and z_o , and these formulas cover all values of z and z_o . However the fact that p is given by several different formulas means that none of them is uniform. Furthermore the eigenvalues and eigenfunctions may be inaccurate for the smallest values of m , when the two turning points are close together.

To obtain a completely uniform asymptotic form for p we must obtain a completely uniform asymptotic form for $\phi_m(z)$. At the same time this will improve the accuracy of the lowest eigenfunctions. An alternative method for improving the accuracy of these eigenfunctions is to introduce a different representation for ϕ_m , involving Weber or parabolic cylinder functions, in a boundary layer around the two nearby turning points. But since a completely uniform form for ϕ_m can be constructed with parabolic cylinder functions, we shall construct it instead.

The uniform asymptotic form of $\phi_m(z)$ just referred to is derived in Appendix 5.2C. It is

5.24
$$\phi_m(z) \sim [S_z(z)]^{-1/2} \Big\{ U[-k/2,(2k)^{1/2}S(z)] - U[-k/2,(2k)^{1/2}S(0)] V[-k/2,(2k)^{1/2}S(z)] / V[-k/2,(2k)^{1/2}S(0)] \Big\} .$$

Here U and V are parabolic cylinder functions, defined in Abramowitz and Stegun [4] page 687, while S(z) is defined by

5.25 $$\frac{S}{2}[1-S^2]^{1/2} + \frac{1}{2}\sin^{-1}S + \frac{\pi}{4} = \int_{z_m}^{z} [n^2(z')-a_m^2]^{1/2}dz' \ .$$

The eigenvalue a_m is the m-th root of the equation

5.26 $$U'[-k/2,(2k)^{1/2}S(-h)]V[-k/2,(2k)^{1/2}S(0)]$$
$$-U[-k/2,(2k)^{1/2}S(0)]V'[-k/2,(2h)^{1/2}S(-h)] = 0 \ .$$

In (5.26) S is defined by (5.25) with a_m replaced by a and z_m replaced by z(a) , the root of the equation n(z) = a .

By using (5.24) for $\phi_m(z)$ in (5.1), together with (5.11) for $H_o^{(1)}$, we obtain a uniform asymptotic form of the modal representation of p(r,z) . From it p can be calculated for all values of z and z_o . Furthermore by expanding the parabolic cylinder functions in it asymptotically, we can recover from this the other asymptotic forms of the modal representation given in this sub-section.

5.2A Appendix

We shall now obtain an asymptotic form of the eigenfunctions $\phi_m(z)$, valid for k large, by the WKB method. To do so we shall first find the asymptotic forms of the upgoing and downgoing waves U and D , defined in sub-section 4.4, because we shall need them later. Then we shall use these forms to find $\phi_m(z)$ asymptotically. The differential equation satisfied by U, D and ϕ_m is (4.3), which is

5.27 $$\phi_{zz} + k^2[n^2(z)-a^2]\phi = 0 \ .$$

We seek a solution ϕ which, for k large, is asymptotically of the form

5.28 $$\phi(z) \sim e^{ikS(z)} \sum_{j=0}^{\infty} (ik)^{-j} A_j(z) .$$

Substituting (5.28) into (5.27) and equating the coefficient of each power of k^{-1} to zero yields

5.29 $$S_z^2 = n^2(z) - a^2 ,$$

5.30 $$2S_z(A_j)_z + S_{zz}A_j = -(A_{j-1})_{zz} , \quad j = 0,1,\cdots; \ A_{-1} \equiv 0 .$$

The solutions of (5.29) which vanish at z_o are

5.31 $$S(z) = \pm \int_{z_o}^{z} (n^2-a^2)^{1/2} dz' .$$

The general solution of (5.30) is, with b_j an arbitrary constant,

5.32 $$A_j(z) = [n^2(z)-a^2]^{-1/4} \left\{ b_j - \frac{1}{2} \int_{z_o}^{z} [n^2(z')-a^2]^{-1/4} [A_{j-1}(z')]_{z'z'} dz' \right\} .$$

Upon using (5.32) and (5.31) with either sign in (5.28), we obtain the asymptotic expansions of two solutions of (5.27). These expansions are valid provided $n^2(z) - a^2$ does not vanish. This is the case if $0 \leq a < n(-h)$. The two solutions are just $U(ka,z)$ and $D(ka,z)$. If we set $b_o = 1$, their leading terms are

$$5.33 \quad U(ka,z) \sim [n^2(z)-a^2]^{-1/4} e^{ik\int_{z_o}^{z} (n^2-a^2)^{1/2}dz'} \quad , \quad -h \le z \le 0 \ ,$$

$$5.34 \quad D(ka,z) \sim [n^2(z)-a^2]^{-1/4} e^{-ik\int_{z_o}^{z} (n^2-a^2)^{1/2}dz'} \quad , \quad -h \le z \le 0 \ .$$

Then by using (5.33) and (5.34) in (4.18) and (4.19) we get

$$5.35 \quad R_1(ka) \sim -e^{2ik\int_{z_o}^{0} (n^2-a^2)^{1/2}dz} \ ,$$

$$5.36 \quad R_2(ka) \sim e^{2ik\int_{-h}^{z_o} (n^2-a^2)^{1/2}dz} \ .$$

To get ϕ_m we form a linear combination of U and D which asymptotically satisfies the boundary condition $\phi_z(-h) = 0$. This can be written as

$$5.37 \quad \phi_m(z) \sim [n^2(z)-a_m^2]^{-1/4} \cos\left[k\int_{-h}^{z} (n^2-a_m^2)^{1/2}dz'\right] \ .$$

Here we have set $a = a_m$, where a_m is determined by the condition $\phi(0) = 0$. This condition yields the eigenvalue equation

$$5.38 \quad k\int_{-h}^{0} (n^2-a_m^2)dz = (m + \tfrac{1}{2})\pi \ , \quad m = 0,1,2,\cdots \ .$$

Let us now consider a in the interval $n(-h) < a < n(0)$, in which case there is exactly one turning point $z(a)$ which satisfies $n[z(a)] = a$. Then U and D are of the form (5.33) and (5.34) in the interval $z(a) < z \le 0$, and it is convenient to replace the lower limit of integration z_o by $z(a)$ in those equations. For $z < z(a)$, U and D are linear combinations of the two solutions corresponding to the two signs in (5.31). To find the appropriate linear combination we use the WKB connection formula, which is (Morse and Feshbach [5] Vol. II, page 1097)

$$5.39 \quad (n^2-a^2)^{-1/4}\left[A \cos\left\{k \int_{z(a)}^{z} (n^2-a^2)^{1/2} dz - \frac{5}{12}\pi\right\} + B \cos\left\{k \int_{z(a)}^{z} (n^2-a^2)^{1/2} dz - \frac{\pi}{12}\right\}\right], z>z(a)$$

$$\longleftrightarrow \quad 1/2(a^2-n^2)^{-1/4}\, e^{-i\pi/4}\left[(B-A)e^{k\int_z^{z(a)} (a^2-n^2)^{1/2} dz} + (A\, e^{i\pi/6} + B\, e^{-i\pi/6})e^{-k\int_z^{z(a)} (a^2-n^2)^{1/2} dz}\right], \quad z<z(a).$$

This relation holds when $n^2 - a^2 > 0$ for $z > z(a)$ and $n^2 - a^2 < 0$ for $z < z(a)$, as in the present case. Here A and B are arbitrary constants.

We now use (5.33) and (5.34) for U and D in the interval $z > z(a)$, with z_o replaced by $z(a)$. Then we use (5.39) to find them for $z < z(a)$. In this way we obtain

$$5.40 \quad U \sim (n^2-a^2)^{-1/4}\, e^{ik\int_{z(a)}^{z} (n^2-a^2)^{1/2} dz} \quad , \quad z(a) < z \le 0 ,$$

$$\sim -i(a^2-n^2)^{-1/4} e^{k \int_z^{z(a)} (a^2-n^2)^{1/2} dz} \quad , \quad -h \leq z < z(a) \ ,$$

5.41 $$D \sim (n^2-a^2)^{-1/4} e^{-ik \int_{z(a)}^{z} (n^2-a^2)^{1/2} dz} \quad , \quad z(a) < z \leq 0 \ ,$$

$$\sim (a^2-n^2)^{-1/4} \left\{ e^{k \int_z^{z(a)} (a^2-n^2)^{1/2} dz} - i\, e^{-k \int_z^{z(a)} (a^2-n^2) dz} \right\} , \quad -h \leq z < z(a) \ .$$

Then (4.18) and (4.19) yield

5.42 $$R_1 \sim -e^{2ik \int_{z(a)}^{0} (n^2-a^2)^{1/2} dz} \ ,$$

5.43 $$R_2 \sim e^{-i\pi/2} + e^{-2k \int_{-h}^{z(a)} (a^2-n^2)^{1/2} dz} \sim e^{-i\pi/2} \ .$$

By forming a linear combination of U and D to satisfy $\phi_z(-h) = 0$ asymptotically, setting $a = a_m$ and $z(a_m) = z_m$, we get

5.44 $$\phi_m(z) \sim (n^2-a^2)^{-1/4} \cos\left[k \int_{z_m}^{z} (n^2-a_m^2)^{1/2} dz - \frac{\pi}{4}\right] , \quad z_m < z \leq 0$$

$$\sim (a_m^2-n^2)^{-1/4} e^{-k\int_z^{z_m} (a_m^2-n^2)^{1/2}dz - i\pi/4} , \quad h \le z < z_m .$$

The condition $\phi_m(0) = 0$ leads to the eigenvalue equation

5.45
$$k\int_{z_m}^{0} (n^2-a_m^2)^{1/2}dz = (m+\tfrac{3}{4})\pi , \quad m = 0,1,2,\cdots .$$

Next we shall consider a in the interval $n(0) < a < n_{max}$. Then there are two turning points $z(a)$ and $z'(a)$ satisfying $n[z(a)] = n[z'(a)] = a$, with $n^2 - a^2 > 0$ for $z(a) < z < z'(a)$. Again U and D are of the form (5.33) and (5.34) with z_o replaced by $z(a)$ in the interval $z(a) < z < z'(a)$. To find them for $z < z(a)$ we use the connection formula (5.29). To find them for $z > z'(a)$ we must first modify the connection formula by replacing $z(a)$ by $z'(a)$ and then interchanging $z'(a)$ with z because $n^2 - a^2 > 0$ for $z < z'(a)$. Then we use the modified formula. In this way we obtain

5.46
$$U \sim e^{ik\int_{z(a)}^{z'(a)} (n^2-a^2)^{1/2}dz} (a^2-n^2)^{-1/4}$$

$$\times\left[e^{k\int_{z'(a)}^{z} (a^2-n^2)^{1/2}dz} - i\, e^{-k\int_{z'(a)}^{z} (a^2-n^2)^{1/2}dz}\right],$$

$$z'(a) < z \le 0 ,$$

$$\sim (n^2-a^2)^{-1/4} e^{ik\int_{z(a)}^{z} (n^2-a^2)^{1/2}dz} , \quad z(a) < z < z'(a) ,$$

$$\sim -i(a^2-n^2)^{-1/4}\, e^{k\int_z^{z(a)} (a^2-n^2)^{1/2}dz} \,, \qquad -h \le z < z(a) \,,$$

5.47 $$D \sim -i(a^2-n^2)^{-1/4}\, e^{-ik\int_{z(a)}^{z'(a)} (n^2-a^2)^{1/2}dz \,+\, k\int_{z'(a)}^{z} (a^2-n^2)^{1/2}dz} \,,$$

$$z'(a) < z \le 0 \,,$$

$$\sim (n^2-a^2)^{-1/4}\, e^{-ik\int_{z(a)}^{z} (n^2-a^2)^{1/2}dz} \,, \qquad z(a) < z < z'(a) \,,$$

$$\sim (a^2-n^2)^{-1/4}\left\{ e^{k\int_z^{z(a)} (a^2-n^2)^{1/2}dz} - i\, e^{-k\int_z^{z(a)} (a^2-n^2)^{1/2}dz} \right\},$$

$$-h \le z < z(a) \,.$$

Now (4.18) and (4.19) yield

5.48 $$R_1 \sim e^{2ik\int_{z(a)}^{z'(a)} (n^2-a^2)^{1/2}dz} \left\{ e^{-i\pi/2} - e^{-2k\int_{z'(a)}^{0} (a^2-n^2)^{1/2}dz} \right\}$$

$$\sim e^{-i\pi/2 \,+\, 2ik\int_{z(a)}^{z'(a)} (n^2-a^2)^{1/2}dz}$$

5.49 $$R_2 \sim e^{-i\pi/2} - e^{-2k\int_{-h}^{z(a)} (a^2-n^2)^{1/2}dz} \sim e^{-i\pi/2} \,.$$

To get $\phi_m(z)$ we again form a linear combination of U and D to satisfy $\phi_z(-h) = 0$ and $\phi(0) = 0$ asymptotically. Then writing a_m instead of a , $z(a_m) = z_m$ and $z'_m = z'(a_m)$ we get the following result for ϕ_m and the eigenvalue equation:

$$5.50 \quad \phi_m(z) \sim -i(a^2-n^2)^{-1/4} e^{ik\int_{z_m}^{z'_m} (n^2-a_m^2)^{1/2}dz - k\int_{z'_m}^{z} (a_m^2-n^2)^{1/2}dz - i\pi/4}, \qquad z'_m < z \leq 0 ,$$

$$\sim (n^2-a_m^2)^{-1/4} \cos\left[k\int_{z_m}^{z}(n^2-a_m^2)^{1/2}dz - \frac{\pi}{4}\right], \quad z_m < z < z'_m,$$

$$\sim (a_m^2-n^2)^{-1/4} e^{-k\int_z^{z_m}(a_m^2-n^2)^{1/2}dz - \frac{\pi}{4}}, \quad -h \leq z < z_m .$$

$$5.51 \quad k\int_{z_m}^{z'_m}(n^2-a_m^2)^{1/2}dz = (m+\tfrac{1}{2})\pi , \quad m = 0,1,2,\cdots .$$

Finally we treat $a > n_{max}$, in which case there are no turning points and no eigenvalues or eigenfunctions. Then (5.33) - (5.36) are valid, but U and D are exponential and not oscillatory.

5.2B <u>Appendix</u>

The asymptotic form (5.44) of the eigenfunction ϕ_m, which holds when there is one turning point z(a) , is not uniform. It consists of two different formulas, neither of which is valid at the turning point. To obtain a uniformly valid asymptotic expansion of ϕ_m we seek ϕ in the form

$$5.52 \qquad \phi(z) \sim W[-k^{2/3}S(z)] \sum_{j=0}^{\infty} k^{-2j}B_j(z) + k^{-4/3}W'[-k^{2/3}S(z)] \sum_{j=0}^{\infty} k^{-2j}C_j(z) \ .$$

Here S, B_j and C_j are to be determined, while W(t) is a solution of the Airy equation

$$5.53 \qquad W''(t) - tW = 0 \ .$$

The form (5.52) is a special case of that given by Lynn and Keller [6] and is esentially due to Langer [7] .

We now substitute (5.52) into the equation (5.27) satisfied by ϕ , and equate to zero the coefficient of each power of k . The first two powers yield

$$5.54 \qquad SS_z^2 - n^2 + a^2 = 0 \ ,$$

$$5.55 \qquad 2S_z(B_o)_z + B_oS_{zz} = 0 \ .$$

The real solution of (5.54) which vanishes at z(a) is given by (5.15) with z(a) replaced by z_m . Then the solution of (5.55) is, apart from an arbitrary constant factor,

$$5.56 \qquad B_o(z) = [S_z(z)]^{-1/2} \ .$$

For W we write a linear combination of the Airy function Ai and Bi , and then the leading term in (5.52) becomes

5.57 $$\phi(z) \sim [S_z(z)]^{-1/2} \left\{ Ai[-k^{2/3}S(z)] + c\ Bi[-k^{2/3}S(z)] \right\} .$$

When the constant c is chosen to make $\phi_z(-h) \sim 0$, (5.57) becomes (5.13). Then the condition $\phi(0) \sim 0$ yields the eigenvalue equation (5.14).

5.2C Appendix

When there are two turning points, the asymptotic form of ϕ_m is given by (5.13) and (5.18) for different ranges of z . We can obtain a uniform expansion of ϕ in this case by seeking ϕ in the form

5.58 $$\phi(z) \sim W[(2k)^{1/2}S(z)] \sum_{j=0}^{\infty} k^{-j}B_j(z) + k^{-3/2}W'[(2k)^{1/2}S(z)] \sum_{j=0}^{\infty} k^{-j}C_j(z) .$$

Again S, B_j and C_j are to be found while W(t) is a solution of the parabolic cylinder equation

5.59 $$W''(t) + \frac{1}{4}(2k-t^2)W(t) = 0 .$$

Upon substituting (5.58) into (5.27), and equating coefficients of the two highest powers of k , we get

5.60 $$S_z^2(S^2-1) + n^2 - a^2 = 0 ,$$

5.61 $$2S_z(B_o)_z + B_oS_{zz} = 0 .$$

The solution of (5.60) is given by (5.25) in which z(a) is replaced by z_m , and the solution of (5.61) is, apart from a constant factor,

$$B_o(z) = [S_z(z)]^{-1/2} . \tag{5.62}$$

The solution W of (5.59) can be written in terms of the functions U and V defined in Abramowitz and Stegun [4] page 687, and then the leading term in (5.58) becomes

$$\phi(z) \sim [S_z(z)]^{-1/2}\left\{U[-k/2,(2k)^{1/2}S(z)] + c\ V[-k/2,(2k)^{1/2}S(z)]\right\} . \tag{5.63}$$

The constant c must be chosen to make $\phi(0) \sim 0$, and then (5.63) becomes (5.24). Then the condition $\phi_z(-h) = 0$ yields the eigenvalue equation (5.26).

5.3 Asymptotic form of the Hankel transform representation

We shall now obtain the asymptotic form of the Hankel transform representation (4.14) of p(r,z) . To do so we shall just replace the functions in the integrand of (4.14) by their WKB asymptotic forms, which can be obtained from the results of Appendix 5.2A. In sub-section 5.5 we shall convert this asymptotic form into a series of integrals, which is the same as the asymptotic form of the multiple scattering representation to be obtained in sub-section 5.4. In sub-section 5.4.1 we shall evaluate these integrals asymptotically by the method of stationary phase. The resulting series is the same as that which will be obtained by the ray method in section 6.

The functions $\tilde{p}_1$ and $\tilde{p}_2$, which occur in (4.14), are given by (4.23) and (4.24) in terms of U, D, R_1 and R_2 . The latter functions are given asymptotically, in Appendix 5.2A, by different formulas in four different ranges of a . Therefore we must divide the range of integration over a in (4.14) into these four ranges, and use the appropriate asymptotic forms in each range. Thus let I_1 be the interval

$0 \le a < n(-h)$, I_2 the interval $n(-h) < a < n(0)$, I_3 the interval $n(0) < a < n_{max}$ and I_4 the interval $a > n_{max}$. Then we can write (4.14) in the form

5.64 $$p(r,z) = \sum_{i=1}^{4} p^{(i)}(r,z)$$

where

5.65 $$p^{(i)}(r,z) = \frac{k^2}{2\pi} \int_{I_i} J_o(kar) \frac{\tilde{p}_1(ka,z_>)\tilde{p}_2(ka,z_<)}{W(ka)} \, ada \, .$$

In (5.65) we use for J_o its asymptotic form

5.66 $$J_o(kar) \sim (2\pi kar)^{-1/2}(e^{ikar-i\pi/4}+e^{-ikar+i\pi/4}) \, .$$

We also use (4.23) and (4.24) for p_1 and p_2 with U, D, R_1 and R_2 given by those formulas in Appendix 5.2A which are valid in I_i. In this way we obtain the following four results:

5.67 $$p^{(1)}(r,z) \sim \left[\frac{k}{2\pi^3 r}\right]^{1/2} \int_0^{n(-h)} \cos(kar - \tfrac{\pi}{4}) \frac{\sin\left[k \int_{z_>}^{0} (n^2-a^2)^{1/2} dz\right]}{[n^2(z)-a^2]^{1/4}[n^2(z_o)-a^2]^{1/4}}$$

$$\times \frac{\cos\left[k \int_{-h}^{z_<} (n^2-a^2)^{1/2} dz\right]}{\cos\left[k \int_{-h}^{0} (n^2-a^2)^{1/2} dz\right]} a^{1/2} da \quad , \quad -h \le z \le 0 \, .$$

5.68 $$p^{(2)}(r,z) \sim \left[\frac{k}{2\pi^3 r}\right]^{1/2} \int_{n(-h)}^{n(0)} \cos(kar - \frac{\pi}{4}) \frac{\sin\left[k \int_{z_>}^{0} (n^2-a^2)^{1/2} dz\right]}{[n^2(z)-a^2]^{1/4}[n^2(z_o)-a^2]^{1/4}}$$

$$\times \frac{\cos\left[k \int_{z(a)}^{z_<} (n^2-a^2)^{1/2} dz - \frac{\pi}{4}\right]}{\cos\left[k \int_{z(a)}^{o} (n^2-a^2)^{1/2} dz - \frac{\pi}{4}\right]} a^{1/2} da \quad , \; z(a)<z \leq 0$$

$$\sim 0 \quad , \; -h \leq z < z(a)$$

5.69 $$p^{(3)}(r,z) \sim 0 \quad , \; z'(a) < z \leq 0 ,$$

$$\sim \left[\frac{k}{2\pi^3 r}\right]^{1/2} \int_{n(0)}^{n_{max}} \frac{\cos(kar - \frac{\pi}{4})\cos\left[k \int_{z_>}^{z'(a)} (n^2-a^2)^{1/2} dz - \frac{\pi}{4}\right]}{[n^2(z)-a^2]^{1/4}[n^2(z_o)-a^2]^{1/4}}$$

$$\times \frac{\cos\left[k \int_{z(a)}^{z_<} (n^2-a^2)^{1/2} dz - \frac{\pi}{4}\right]}{\cos\left[k \int_{z(a)}^{z'(a)} (n^2-a^2)^{1/2} dz\right]} a^{1/2} da \quad , \; z(a) < z < z'(a),$$

$$\sim 0 \quad , \; -h \leq z < z(a) .$$

5.70 $$p^{(4)}(r,z) \sim 0 \quad , \; -h \leq z \leq 0 .$$

Finally the asymptotic form of p is given by (5.64) with the $p^{(i)}$ given by (5.67)-(5.70). From (5.70) we see that $p^{(4)}$ is exponentially small, so it can be omitted from the sum for p, which then consists of just three integrals.

In sub-section 5.5 we shall show how this result can be converted into the asymptotic forms of the normal mode and the multiple scattering representations, and also how it can be obtained from them.

5.4 Asymptotic form of the multiple scattering representation

The multiple scattering representation (4.22) of $p(r,z)$ can be expanded asymptotically by using (5.66) for $J_o(kar)$ and the formulas for U, D, R_1 and R_2 given in Appendix 5.2A. There are four sets of these formulas corresponding to four different intervals of the parameter a. Therefore we must first split each integral in (4.22) into a sum of four integrals over these four intervals, and then use the appropriate asymptotic forms from Appendix 5.2A in each integral. After proceeding in this way, we can write the asymptotic form of (4.22) in the form

5.71 $$p(r,z) = \sum_{i=1}^{4} \sum_{\pm} p_i^{\pm}(r,z) = \sum_{i=1}^{4} \sum_{j=1}^{4} \sum_{\pm} p_{ij}^{\pm}(r,z) .$$

The function $p_{ij}^{\pm}$ in (5.71) is defined by

5.72 $$p_{ij}^{\pm}(r,z) \sim \left[\frac{k}{32\pi^3 r}\right]^{1/2} \sum_{m=0}^{\infty} (-1)^m a_{ijm} \int_{I_i} \frac{e^{\pm i\pi/4 + ikS_{ij}^{\pm}(r,z,a;m)}}{[n^2(z)-a^2]^{1/4}[n^2(z_0)-a^2]^{1/4}} a^{1/2} da ,$$

$i = 1,2,3,\ z \text{ in } R_i ,$

$$p_{ij}^{\pm}(r,z) \sim 0 , \quad z \text{ not in } R_i$$

$$p_{4j}^{\pm}(r,z) \sim 0 , \ -h \leq z \leq 0 .$$

The coefficient a_{ijm}, the range R_i and the interval I_i are given in the following table:

5.73

i	a_{ijm}	R_i	I_i
1	$(-1)^{j-1}$	$[-h, 0]$	$[0, n(-h)]$
2	$(-e^{i\pi/2})^{j-1}\, e^{-im\pi/2}$	$(z(a), 0]$	$[n(-h), n(0)]$
3	$(-1)^{j-1}$ for $j = 1,2$; $e^{-i\pi/2}$ for $j = 3,4$.	$(z(a), z'(a))$	$[n(0), n_{max}]$

The function $S_{ij}^{\pm}$ in (5.72) is defined as follows:

$$5.74 \quad S_{ij}^{\pm} = \pm ar + \left\{(-1)^{j-1}\int_{z_<}^{z_>} + 2(m+j-1)\int_{-h}^{0}\right\}(n^2-a^2)^{1/2}dz\ , \quad j = 1,2\ .$$

For j = 3 or 4 , $S_{1j}^{\pm}$ is given by (5.74) with j replaced by j-2 and $\int_{z_<}^{z_>}$ replaced by $\int_{z_0}^{0} + \int_{z}^{0}$. $S_{2j}^{\pm}$ is obtained from $S_{1j}^{\pm}$ by replacing the lower limit of integration -h by z(a) . $S_{3j}^{\pm}$ is obtained from $S_{1j}^{\pm}$ by replacing the limits of integration -h and 0 by z(a) and z'(a) respectively.

Thus the asymptotic form of the multiple scattering representation of p(r,z) is given by (5.71) with the $p_{ij}^{\pm}$ given by (5.72). Each term in this sum for p has an interpretation as a multiply scattered wave, just like the terms in (4.22). In the next sub-section we shall simplify (5.72) further by evaluating all the integrals asymptotically for k large, using the method of stationary phase. The result has an interpretation in terms of rays. It will be rederived directly by the ray method in section 6.

In sub-section 5.5 we shall show how (5.71) can be obtained from (5.64), the asymptotic form of the Hankel transform representation, and also how (5.71) can be summed to yield (5.64).

5.4.1 Explicit asymptotic form of the multiple scattering representation

Now we shall evaluate the integrals in (5.71) asymptotically for k large. This will yield a simpler asymptotic form of p(r,z). To evaluate them we shall use the method of stationary phase, which is explained, for example, in Erdelyi [8] . First we note from (5.72) that $p_{4j}^{\pm} \sim 0$. Next we find that the phase function S_{ij}^{-} has no stationary points, so $p_{ij}^{-} \sim 0$. Therefore (5.71) can be written as

$$5.75 \quad p(r,z) \sim \sum_{i=1}^{3}\sum_{j=1}^{4} p_{ij}^{+}\ .$$

Now we shall evaluate the integral (5.72) for $p_{ij}^{\pm}$ by stationary phase.

From (5.74) we find that the stationary points of the phase function S_{1j}^{+}, $j = 1,2$ satisfy the equation

$$5.76 \qquad r = a\left\{(-1)^{j-1}\int_{z_<}^{z_>} + 2(m+j-1)\int_{-h}^{0}\right\}(n^2-a^2)^{-1/2}dz \ , \ j = 1,2; \ m = 0,1,2,\cdots .$$

Let $a = b_m$ be the root of (5.76) for $m = 0,1,2,\cdots$. The second derivative of S_{1j}^{+} with respect to a, evaluated at $a = b_m$, occurs in the stationary phase result. From (5.74) we obtain for it

$$5.77 \qquad \frac{\partial^2 S_{1j}^{+}}{\partial a^2} = -\left\{(-1)^{j-1}\int_{z_<}^{z_>} + 2(m+j-1)\int_{-h}^{0}\right\} n^2(n^2-b_m^2)^{-3/2}dz \ , \quad j = 1,2 .$$

The stationary points of all the other S_{1j}^{+} satisfy the equations obtained from (5.76) by the replacements described after (5.74). Also $\partial^2 S_{1j}^{+}/\partial a^2$, $j = 3,4$, is given by (5.77) with the same replacements. However the second derivatives of the other S_{ij}^{+} cannot be obtained from (5.77) because for them one or more of the limits of integration are zeros of n^2-a^2, and that would lead to singular integrals.

To calculate the second derivative of S_{2j}^{+}, $j = 1,2$ we first replace $-h$ by $z(a)$ in (5.74) and then differentiate once to obtain

$$5.78 \qquad \frac{\partial S_{2j}^{+}}{\partial a} = r - a\left\{(-1)^{j-1}\int_{z_<}^{z_>} + 2(m+j-1)\int_{z(a)}^{0}\right\}(n^2-a^2)^{-1/2}dz \ , \quad j = 1,2 .$$

We now add to and subtract from the integrand the quantity $n^2(n^2-a^2)^{-1/2}/a$. In the subtracted term we write n^2 as $(n/n')\,nn'$ and integrate by parts to obtain

$$5.79 \quad \frac{\partial S_{2j}^{+}}{\partial a} = r+(-1)^{j-1}\left\{ \frac{\rho(z_<)}{a}\,[n^2(z_<)-a^2]^{1/2} - \frac{\rho(z_>)}{a}\,[n^2(z_>)-a^2]^{1/2}\right\}$$

$$-2(m+j-1)\,\frac{\rho(0)}{a}\,[n^2(0)-a^2]^{1/2} + \frac{1}{a}\left\{(-1)^{j-1}\int_{z_<}^{z_>} +2(m+j-1)\int_{z(a)}^{0}\right\}(1+\rho')(n^2-a^2)^{1/2}\,dz,$$

$$j = 1,2 .$$

Here

$$5.80 \quad \rho(z) = n(z)/n'(z) .$$

Differentiating (5.79) and setting $a = b_m$ yields

$$5.81 \quad \frac{\partial^2 S_{2j}^{+}}{\partial a^2} = (-1)^j\left\{\frac{\rho(z_<)n^2(z_<)}{b_m^2[n^2(z_<)-b_m^2]^{1/2}} + \frac{\rho(z_>)n^2(z_>)}{b_m^2[n^2(z_>)-b_m^2]^{1/2}}\right\}$$

$$+\,2(m+j-1)\,\frac{\rho(0)n^2(0)}{b_m^2[n^2(0)-b_m^2]^{1/2}} + \frac{1}{b_m^2}\left\{(-1)^j\int_{z_<}^{z_>} +(2m+j-1)\int_{z(b_m)}^{0}\right\}\frac{(1+\rho')n^2}{(n^2-b_m^2)^{3/2}}\,dz ,$$

$$j = 1,2 .$$

Similarly we find

$$5.82 \quad \frac{\partial^2 S_{23}^{+}}{\partial a^2} = -\,\frac{\rho(z_0)n^2(z_0)}{b_m^2[n^2(z_0)-b_m^2]^{1/2}} - \frac{\rho(z)n^2(z)}{b_m^2[n^2(z)-b_m^2]^{1/2}}$$

$$+\,(2m+2)\,\frac{\rho(0)n^2(0)}{b_m^2[n^2(0)-b_m^2]^{1/2}} - \frac{1}{b_m^2}\left\{\int_{z_0}^{0} + \int_{z}^{0} + 2m\int_{z(b_m)}^{0}\right\}\frac{(1+\rho')n^2}{(n^2-b_m^2)^{1/2}}\,dz .$$

$$5.83 \quad \frac{\partial^2 S_{24}^+}{\partial a^2} = \frac{\rho(z_0)n^2(z_0)}{b_m^2[n^2(z_0)-b_m^2]^{1/2}} + \frac{\rho(z)n^2(z)}{b_m^2[n^2(z)-b_m^2]^{1/2}} + 2m\,\frac{\rho(0)n^2(0)}{b_m^2[n^2(0)-b_m^2]^{1/2}}$$

$$-\frac{1}{b_m^2}\left\{-\int_{z_0}^{0} - \int_{z}^{0} + (2m+2)\int_{z(b_m)}^{0}\right\}\frac{(1+\rho')n^2}{(n^2-b_m^2)^{1/2}}\,dz$$

We also find that $\partial^2 S_{3j}^+/\partial a^2$ is given by the same expression as $\partial^2 S_{2j}^+/\partial a^2$ with $\rho(0)$ replaced by 0 and with the upper limit 0 replaced by $z'(b_m)$.

With these preliminary calculations completed, we can apply the stationary phase formula to p_{ij}^+ given by (5.72). The result is

$$5.84 \quad p_{ij}^+(r,z) \sim \frac{1}{4\pi}\sum_{m=0}^{\infty}(-1)^m a_{ijm}\, b_m^{1/2}\left[-r\,\frac{\partial^2 S_{ij}^+}{\partial a^2}\right]^{-1/2}$$

$$\left[n^2(z_0)-b_m^2\right]^{-1/4}\left[n^2(z)-b_m^2\right]^{-1/4} e^{ikS_{ij}^+} ,$$

$$i = 1,2,3;\ z \text{ in } R_i \ .$$

$$\sim 0 \qquad\qquad z \text{ not in } R_i$$

By using (5.84) in (5.75) we obtain the desired explicit asymptotic form of the multiple scattering representation of $p(r,z)$.

5.5 Connections between the asymptotic forms of the representations

We have now obtained asymptotic forms of the modal, Hankel transform and multiple scattering representations of p , as well as a simpler asymptotic form of the latter. These four asymptotic forms are indicated in Figure 4. We shall now show how these different asymptotic forms can be transformed into one another. The transformations are also indicated in the figure.

Let us begin with (5.64), the asymptotic form of the Hankel transform representation, which expresses p as a sum of the four $p^{(i)}$. Each of the $p^{(i)}$ contains a cosine factor in the denominator of the integrand. We first rewrite these factors by using the following expansion, which is obtained by using the binomial theorem:

$$\frac{1}{\cos x} = \frac{2}{e^{-ix}(1+e^{2ix})} = \sum_{j=0}^{\infty} (-1)^j e^{(2j+1)x} .$$

We also write the trigonometric functions in the numerators of the integrands in terms of exponentials. In this way (5.64) becomes transformed exactly into (5.71), the asymptotic form of the multiple scattering representation. By reversing these steps, we can transform the asymptotic form (5.71) of the multiple scattering representation into the asymptotic form (5.64) of the Hankel transform representation. These transformations are indicated by the line labeled "binomial expansion" in figure 4.

We shall now show how the asymptotic Hankel transform representation (5.64), with the $p^{(i)}$ given by (5.67)-(5.70) can be converted into the asymptotic normal mode representation (5.12). First, as we have shown in sub-sections 3.5 and 4.5, we can write the asymptotic Hankel transform representation in the form

5.85 $$p(r,z) \sim \left[\frac{k}{2\pi}\right]^{3/2} r^{-1/2} \lim_{R\to\infty} \int_{C_1+C_2+\Gamma_R} e^{i(kar-\pi/4)}$$

$$\cdot\ \tilde{p}_1(ka,z_>)\tilde{p}_2(ka,z_<)W^{-1}(ka)^{1/2} a\,da .$$

The contour $C_1 + C_2 + \Gamma_R$ is shown in Figure 3. The functions $\tilde{p}_1$, $\tilde{p}_2$ and W(ka) are given in sub-section 5.3 by different representations for different ranges of the real part of a.

Let $a = a_m$, $m = 0,1,2,\cdots$, be the roots of the equation $W(ka) = 0$ which lie in the upper half of the a-plane. Those real roots which lie between $(0,n(-h))$, $(n(-h), n(0))$ and $(n(0), n_{max})$ are the zeros of the Wronskian $W(ka)$ of $\tilde{p}_1$, $\tilde{p}_2$ given by (4.23), (4.24) with U, D, R_1, and R_2 given by those formulas in Appendix 5.2A which are valid in I_1, I_2 and I_3 respectively. They are exactly the values given by (5.3), (5.6) and (5.9).

We now compute the integral in (5.85) by the method of residues and obtain

$$5.86 \quad p(r,z) \sim \left[\frac{k}{2\pi r}\right]^{1/2} \sum_m a_m^{1/2} e^{i(ka_m r+\pi/4)} \frac{\tilde{p}_1(ka_m,z_>)\tilde{p}_2(ka_m,z_<)}{W'(ka_m)} .$$

We next replace $\tilde{p}_1$, $\tilde{p}_2$ and W' by the representation appropriate to each value of a_m. Then we find that (5.86) simplifies to the asymptotic normal mode representation (5.12). By reversing the steps we can obtain the asymptotic Hankel transform representation from the asymptotic normal mode representation. These transformations are indicated by the line labele "residues" in Figure 4.

Next we shall show how to convert the simpler asymptotic form of the multiple scattering representation (5.75) into the asymptotic normal mode representation (5.12) [9]. First we use (5.73), (5.74) and (5.77) in (5.84) for $i = j = 1$ to obtain

$$p_{11}^{+} \sim \sum_{m=0}^{\infty} \frac{b_m^{1/2}}{4\pi} \cdot \frac{e^{ik\left[b_m r + \left\{\int_{z_<}^{z_>} + 2m\int_{-h}^{0}\right\}(n^2-b_m^2)^{1/2}dz \; -im\pi\right]}}{[n^2(z)-b_m^2]^{1/4}[n^2(z_0)-b_m^2]^{1/4}\left[r\left\{\int_{z_<}^{z_>} + 2m\int_{-h}^{0}\right\}n^2(n^2(n^2-b_m^2)^{-3/2}dz\right]^{1/2}} .$$

Here we have used $(-1)^m = e^{-im\pi}$. We now sum this series by Poisson's formula (3.34) to get

$$5.87 \quad p_{11}^{+} \sim \frac{1}{8\pi^2} \sum_{q=-\infty}^{\infty}$$

$$\cdot \int_0^\infty \frac{a_\zeta^{1/2} e^{ik\left[a_\zeta r + \left\{\int_{z_<}^{z_>} + \frac{\zeta}{\pi}\int_{-h}^{0}\right\}(n^2-a_\zeta^2)^{1/2}dz\right] - \frac{i\zeta}{2} - iq\zeta}}{[n^2(z)-a_\zeta^2]^{1/4}[n^2(z_0)-a_\zeta^2]^{1/4}\left[r\left\{\int_{z_<}^{z_>} + \frac{\zeta}{\pi}\int_{-h}^{0}\right\} n^2(n^2-a_\zeta^2)^{-3/2}\right]^{1/2}} d\zeta \, .$$

Here we have used the notation $a_\zeta = b_{\zeta/2\pi} = b_m$. The stationary points of (5.87) are given by the zeros of $\frac{d\Phi}{d\zeta} = 0$ where Φ is the coefficient of ik in the exponent and it is

$$\Phi(a_\zeta,\zeta) = a_\zeta r + \left\{\int_{z_<}^{z_>} + \frac{\zeta}{\pi}\int_{-h}^{0}\right\}(n^2-a_\zeta^2)^{1/2}dz - \frac{\zeta}{2k} - \frac{q\zeta}{k} \, .$$

Since $\frac{\partial\Phi}{\partial a_\zeta} = r - \left\{\int_{z_<}^{z_>} + \frac{\zeta}{\pi}\int_{-h}^{0}\right\} \frac{a_\zeta}{(n^2-a_\zeta^2)^{1/2}} dz = 0$ due to the equation (5.76) for j = 1,

the stationary points are given by $\frac{d\Phi}{d\zeta} = \frac{\partial\Phi}{\partial\zeta} = 0$. This gives

$$5.88 \quad \frac{1}{\pi}\int_{-h}^{0}(n^2-a_\zeta^2)^{1/2}dz - \frac{1}{2k} - \frac{q}{k} = 0 \, .$$

Since the solutions a_ζ depend upon q, this is the eigenvalue relation (5.3) if we designate a_ζ by a_q. We now compute $\frac{d^2\Phi}{d\zeta^2}$ and find

5.89 $$\frac{d^2\Phi}{d\zeta^2} = -\frac{1}{\pi}\left[\int_{-h}^{0} \frac{a_\zeta}{(n^2-a_\zeta^2)^{1/2}}\,dz\right]\frac{da_\zeta}{d\zeta}.$$

In order to find $\frac{da_\zeta}{d\zeta}$, we differentiate the relation given by $\frac{\partial\Phi}{\partial a_\zeta} = 0$ with respect to ζ and obtain

$$\frac{da_\zeta}{d\zeta} = -\frac{1}{\pi}\int_{-h}^{0} a_\zeta\,(n^2-a_\zeta^2)^{-1/2}dz \Bigg/ \left\{\int_{z_<}^{z_>} + \frac{\zeta}{\pi}\int_{-h}^{0}\right\} n^2(n^2-a_\zeta^2)^{-3/2}\,dz\,.$$

By using this equation in (5.89), we get

5.90 $$\frac{d^2\Phi}{d\zeta^2} = \frac{1}{\pi^2}\,\frac{\left[\int_{-h}^{0} a_\zeta(n^2-a_\zeta^2)^{-1/2}dz\right]^2}{\left\{\int_{z_<}^{z_>} + \frac{\zeta}{\pi}\int_{-h}^{0}\right\} n^2(n^2-a_\zeta^2)^{-3/2}dz}\,dz\,.$$

By using (5.87) and (5.90), we find that the stationary point contribution in (5.87) yields

5.91 $$p_{11}^{+} \sim (32\pi kr)^{-1/2}\sum_q \frac{e^{ik[a_q r + \int_{z_<}^{z_>}(n^2-a_q^2)^{1/2}dz] + i^{\pi}/_4}}{[n^2(z)-a_q^2]^{1/4}[n^2(z_0)-a_q^2]^{1/4}\int_{-h}^{0}[a_q/(n^2-a_q^2)]^{1/2}dz}\,.$$

Similarly we can show that

5.92 $$p_{12}^{+} \sim (32\pi kr)^{-1/2}\sum_q \frac{e^{ik[a_q r - \int_{z_<}^{z_>}(n^2-a_q^2)^{1/2}dz] + i^{\pi}/_4}}{[n^2(z)-a_q^2]^{1/4}[n^2(z_0)-a_q^2]^{1/2}\int_{-h}^{0}[a_q/(n^2-a_q^2)]^{1/2}dz}\,,$$

$$5.93 \qquad p_{13}^{+} \sim -(32\pi kr)^{-1/2} \sum_q \frac{e^{ik\left[a_q r + \left\{\int_{z_0}^{0} + \int_{z}^{0}\right\}(n^2-a^2)^{1/2}dz\right] + i\pi/4}}{[n^2(z)-a_q^2]^{1/4}[n^2(z_0)-a_q^2]^{1/4}\int_{-h}^{0}[a_q/(n^2-a_q^2)]^{1/2}dz} \; .$$

$$5.94 \qquad p_{14}^{+} \sim (32\pi kr)^{-1/2} \sum_q \frac{e^{ik[a_q r + \left\{\int_{-h}^{z_0} + \int_{-h}^{z}\right\}(n^2-a_q^2)^{1/2}dz] + i^{\pi}/4}}{[n^2(z)-a^2]^{1/4}[n^2(z_0)-a^2]^{1/4}\int_{-h}^{0}[a_q/(n^2-a_q^2)]^{1/2}dz} \; .$$

Then we add these four representations and after using (5.88) in this sum, we find that it is exactly the second term of the asymptotic modal representation (5.12). Similarly we use Poisson's summation formula (3.34) in $\sum_{j=1}^{4} p_{2j}$ and $\sum_{j=1}^{4} p_{3j}$ and then use the method of stationary phase in these representations. The stationary phase condition yields the eigenvalue relations (5.6) and (5.9). Finally we find that $\sum_{j=1}^{4} (p_{2j} + p_{3j})$ is exactly the same as the first term in the asymptotic modal representation (5.12). Hence we have shown that the asymptotic modal representation (5.12) can be obtained from the simpler asymptotic multiple scattering representation (5.75). This derivation is indicated "Poisson summation" in figure 4.

This completes the derivation of all the connections between the various representations shown in figure 4.

6. The ray representation

6.1 Introduction

The exact and asymptotic representations of p in sections 4 and 5 have been derived for a horizontally stratified ocean of constant depth. We shall now explain how to obtain a representation of p for an unstratified ocean of nonuniform depth. We call it the ray representation because it is based on the rays of geometrical acoustics. Since it is based on rays, it is valid only when the acoustic wavelength is small compared to the scale lengths of the refractive index variations and of the horizontal depth variations.

First we shall describe how to construct the ray representation synthetically, by following a recipe or set of rules. These are the rules of geometrical acoustics, which have a clear physical significance and which provide an intuitively appealing picture of the process of wave propogation. Next we shall show how to obtain the ray representation analytically, by deriving it directly from the reduced wave equation and the appropriate boundary conditions. Finally we shall specialize the ray representation to the horizontally stratified ocean of constant depth. Then it will become exactly the explicit asymptotic form of the multiple scattering representation given in subsection 5.4.1 by (5.75) and (5.84).

6.2 Geometrical construction of the ray representation

To construct the ray representation of $p(\underline{x})$ according to geometrical acoustics, we must carry out the following steps:

1. Determine all the rays from the source point $\underline{x}_0$ to the field point $\underline{x}$. These include the direct ray, the rays refracted any number of times in a sound channel and the rays reflected any number of times at the top and bottom surfaces. To obtain a more complete representation, various kinds of diffracted rays and complex rays may have to be included.
2. Calculate the optical length $S^{(j)}(\underline{x})$ of the j-th ray from $\underline{x}_0$ to $\underline{x}$.
3. Calculate the amplitude $A^{(j)}(\underline{x})$ of the field on the j-th ray at $\underline{x}$. This involves conservation of flux in a ray tube, reflection coefficients at the top and bottom surfaces, change of phase at a caustic, etc.
4. Combine the fields $A^{(j)}(\underline{x})\, e^{ikS^{(j)}(\underline{x})}$ on all the rays through $\underline{x}$ to obtain the ray representation of $p(\underline{x})$ in the form

$$p(\underline{x}) \sim \sum_j A^{(j)}(\underline{x}) e^{ikS^{(j)}(\underline{x})} \quad . \tag{6.1}$$

On a complex ray, the phase $S^{(j)}(x)$ is complex and the corresponding field is evanescent. On a diffracted ray, the amplitude $A^{(j)}(\underline{x})$ is inversely proportional to some fractional power of k, so the corresponding field is weaker than that on an ordinary ray. At a caustic associated with the j-th ray, $A^{(j)}(\underline{x})$ is infinite and a different expression for the field on that ray must be used. Both boundary layer theory and the uniform representation of Kravtsov and Ludwig provide correct expressions for this field.

The j-th term in the sum (6.1) can be interpreted as the leading term in the asymptotic expansion for k large, of the field on the j-th ray. This will be shown by the analytic derivation of the field on the j-th ray in the next sub-section. Furthermore, that derivation will show how to construct further terms in this field.

The representation (6.1) has been used widely to calculate $p(\underline{x})$ in horizontally stratified oceans of constant depth. In this case the amplitude $A^{(j)}(\underline{x})$ can be expressed in a relatively simple form in terms of the refractive index $n(z)$. However (6.1) is not so convenient to use in the more general case of an unstratified ocean of either constant or non-constant depth. This is because of the numerical difficulty of solving the transport equation for the variation of the amplitude along a ray, since this equation involves the divergence of neighboring rays. As a consequence (6.1) has not been used widely in the case of an unstratified ocean. Therefore, the horizontal ray method of chapter III, and the parabolic equation method of chapter V, have been devised for use in the non-stratified case.

6.3 Analytic derivation of the ray representation

In the ray representation (6.1), the pressure p is represented as a sum of terms. Each term consists of a phase factor $e^{ikS(\underline{x})}$ and an amplitude factor $A(\underline{x})$. We consider each such term to be the leading term in an asymptotic expansion of the form

$$6.2 \qquad p(x) \sim e^{ikS(\underline{x})} \sum_{m=0}^{\infty} (ik)^{-m} A_m(\underline{x}) \; .$$

The coefficient $A_0(\underline{x})$ in (6.2) is just the $A(\underline{x})$ which occurs in (6.1), and the other $A_m(\underline{x})$ represent corrections to it. We call the right side of (6.2) a wave. We shall first show how to determine $S(\underline{x})$ and the $A_m(\underline{x})$ so that the wave (6.2) is an asymptotic solution of the reduced wave equation

$$6.3 \qquad \Delta p + k^2 n^2(\underline{x}) p = 0.$$

Then we shall show how the initial values of S and the A_m are determined by the source. Finally we shall form a sum of waves to satisfy the boundary conditions.

We begin by substituting (6.2) into (6.3) and collecting the coefficients of each power of k. Then we equate each such coefficient to zero and obtain the following equations:

$$(\nabla S)^2 = n^2(\underline{x}) \ , \tag{6.4}$$

$$2\nabla S \cdot \nabla A_m + A_m \Delta S = -\Delta A_{m-1} \ , \quad m = 0,1,\cdots, \quad A_{-1} \equiv 0 \ . \tag{6.5}$$

Equation (6.4) is the eiconal equation of geometrical acoustics, from which the phase function S can be determined. Then (6.5) form a recursive system of first order linear partial differential equations from which the A_m can be found successively, starting with m = 0 .

To solve (6.4) we introduce a two parameter family of curved lines, called rays, which are orthogonal to the level surfaces of S. If we denote the parameters by a and ϕ and let σ denote arclength along a ray, then we can write the rays as

$$\underline{x} = \underline{x}(\sigma, a, \phi) \ . \tag{6.6}$$

The orthogonality of the rays and the level surfaces, which are called wavefronts, is expressed by

$$\frac{d\underline{x}}{d\sigma} = \frac{1}{n} \nabla S \ . \tag{6.7}$$

We can eliminate ∇S by differentating (6.7) and using (6.4) to obtain

$$n\frac{d}{d\sigma}\left(n\frac{d\underline{x}}{d\sigma}\right) = \frac{1}{2} \nabla n^2 \ . \tag{6.8}$$

This is a set of three second order ordinary differential equations for $\underline{x}$,called the ray equations.

Now we can write (6.4) as an ordinary differential equation along a ray by using (6.7), which yields

$$\frac{dS}{d\sigma} = n \ . \tag{6.9}$$

The solution of (6.9) is

$$6.10 \qquad S(\sigma) = S(\sigma_0) + \int_{\sigma_0}^{\sigma} n[\underline{x}(\sigma')]d\sigma' .$$

Here $S(\sigma)$ is the value of S at $\underline{x}(\sigma)$, and the parameters a, ϕ have been omitted.

Next by using (6.7), we can write (7.5) as the following ordinary differential equation along a ray

$$6.11 \qquad 2n\frac{dA_m}{d\sigma} + A_m \Delta S = -\Delta A_{m-1} , \qquad m = 0,1,\cdots, \quad A_{-1} \equiv 0 .$$

These equations are called the transport equations. The coefficient ΔS in (6.11) can be expressed in terms of $J(\sigma,a,\phi)$, the Jacobian of the transformation (6.6) from the ray coordinates σ,a,ϕ to the cartesian coordinates $\underline{x}$, defined by

$$6.12 \qquad J = \left| \frac{\partial \underline{x}}{\partial(\sigma,a,\phi)} \right| .$$

In terms of J, ΔS is given by [10]

$$6.13 \qquad \Delta S = \frac{1}{J}\frac{d}{d\sigma}(nJ) .$$

We now substitute (6.13) into (6.11) and then solve (6.11) to obtain

$$6.14 \qquad A_m(\sigma) = \left| \frac{n(\sigma_0)J(\sigma_0)}{n(\sigma)J(\sigma)} \right|^{1/2} A_m(\sigma_0) - \frac{1}{2}\int_{\sigma_0}^{\sigma} \left| \frac{n(\sigma')J(\sigma')}{n(\sigma)J(\sigma)} \right|^{1/2} \Delta A_{m-1}(\sigma')d\sigma' .$$

The results (6.10) and (6.14) determine S and the A_m in terms of the rays and the initial values $S(\sigma_0)$ and $A_m(\sigma_0)$. These initial values are determined by conditions at the initial point of each ray. Thus for example, the direct wave involves rays which start from the point source at $\underline{x}_0$ with phase $S(\underline{x}_0) = 0$. If we set $\sigma_0 = 0$ then the initial conditions for $\underline{x}$ are $\underline{x}(0,a,\phi) = \underline{x}_0$ and $d\underline{x}(0,a,\phi)/d\sigma = \underline{U}(a,\phi)$ where $\underline{U}$ is a unit vector. These conditions and (6.8) determine the direct rays. Then (6.10) with the initial value $S(0,a,\phi) = 0$ determines S. To find the initial values of the A_m, we introduce the source term $-\delta(\underline{x}-\underline{x}_0)$ on the right side of (6.3). Then we find that [11] $A_m(0) = 0$ for $m = 1,2,\cdots$, and

$$6.15 \qquad \lim_{\sigma_0 \to 0} J^{1/2}(\sigma_0) A_0(\sigma_0) = \frac{1}{4\pi} \left[\frac{\tan \alpha}{n(\underline{x}_0)} \right]^{1/2} .$$

Here we have chosen the ray parameter to be $a = n(\underline{x}_0) \sin \alpha$ where α and ϕ are the spherical polar coordinates. By using these results we can determine the direct wave completely.

To satisfy the boundary condition (4.2a) at the top surface, we introduce a top reflected wave, which is also of the form (6.2). We then substitute the sum of the direct or incident wave and the reflected wave into the boundary condition, and equate to zero the coefficient of each power of k. In this way we find that at the boundary the reflected phase is equal to the incident phase and that each A_m in the reflected wave is equal to minus the corresponding A_m in the incident wave. The phase condition leads to the law of reflection for the reflected rays and provides the initial condition for the reflected phase. This initial condition and the initial conditions for the A_m enable us to solve for S and the A_m on the reflected rays. This enables us to construct the top reflected wave completely.

To satisfy the boundary condition on the bottom surface, we introduce a bottom reflected wave. We proceed similarly to substitute the sum of this wave and the incident wave into the boundary condition. In this way we obtain initial conditions for the determination of the bottom reflected wave.

Each wave reflected from one boundary may hit the opposite boundary. Then it leads to a new reflected wave which can be found in the manner described above. In this way an infinite sequence of multiply reflected waves can be obtained.

The family of rays associated with any wave may have an envelope or caustic surface. Then the rays do not cross the caustic but turn away from it. The family of turned rays constitute a new refracted wave which is again of the form (6.2). This refracted wave is given by the same expression as the wave incident upon the caustic, but with a phase change of amount $-\pi/2$.

The pressure $p(\underline{x})$ is given by the sum of all the waves at $\underline{x}$. This includes the direct wave, the waves singly and multiply reflected at the top and bottom surfaces, the waves refracted one or more times, and the waves which are both reflected and refracted any number of times. In addition there may be diffracted and evanescent waves associated with diffracted rays and complex rays, respectively. The sum of the leading terms of all these is just the ray representation (6.1).

6.4 The ray representation for the stratified ocean of constant depth

We shall now apply the method of the preceeding sub-section to the special case of a stratified ocean of constant depth governed by (4.1) and (4.2). In this case the ray equations (6.8) become

$$\frac{d^2x}{d\sigma^2} = 0 \quad , \quad \frac{d^2y}{d\sigma^2} = 0 \, , \quad \frac{d}{d\sigma}\left[\, n(z)\,\frac{dz}{d\sigma}\,\right] = \frac{dn}{dz} \, . \tag{6.16}$$

We multiply the last equation in (6.16) by $n dz/d\sigma$ and integrate to obtain

$$\left(\frac{dz}{d\sigma}\right)^2 = \frac{n^2-a^2}{n^2} \, . \tag{6.17}$$

Here the integration constant is just $a = n(z_0) \sin \alpha$, where α is the initial angle between the ray and the z-axis. From the first two equations in (6.16), it follows that each ray lies in a plane normal to $z = 0$. Since the rays start at the source $\underline{x}_0 = (0,0,z_0)$, each ray lies in a plane $y/x = \tan\theta =$ constant. If we set $r^2=x^2+y^2$, then it follows from this fact, (6.17) and the arclength condition $(d\underline{x}/d\sigma)^2 = 1$ that

$$\left(\frac{dr}{d\sigma}\right)^2 = \frac{a^2}{n^2} \, . \tag{6.18}$$

By combining (6.17) and (6.18) we obtain

$$\frac{dr}{dz} = \frac{\pm a}{(n^2-a^2)^{1/2}} \, . \tag{6.19}$$

Integrating (6.19) with $r = 0$ at $z = z_0$ yields

$$r = \pm \int_{z_0}^{z} \frac{a}{(n^2-a^2)^{1/2}}\, dz = \int_{z_<}^{z_>} \frac{a\,dz}{(n^2-a^2)^{1/2}} \, . \tag{6.20}$$

In order to make $r \geq 0$, we have chosen the plus sign if $z > z_0$ and the minus sign if $z < z_0$. The result (6.20) gives the equation of a direct ray from the source with the ray parameter a .

We next compute the phase S given by (6.10) with $S(\sigma_0) = 0$ and obtain

$$S(r,z) = \int_{\sigma_0}^{\sigma} n[z(\sigma')]d\sigma' = \int_{z_0}^{z} n(z) \frac{d\sigma'}{dz} dz \; . \tag{6.21}$$

By using (6.17) for $d\sigma/dz$ with the appropriate sign, we get from (6.21)

$$S(r,z) = \int_{z_<}^{z_>} \frac{n^2}{(n^2-a^2)^{1/2}} dz \; . \tag{6.22}$$

Subtracting and adding a^2 in the numerator of the integrand, and then using (6.20), enables us to write (6.22) in the form

$$S(r,z) = ar + \int_{z_<}^{z_>} (n^2-a^2)^{1/2} dz \; . \tag{6.23}$$

Now to find the amplitude A_0 we use (6.14) with m=0 and note that $A_{-1}=0$. We evaluate the Jacobian J by introducing the cylindrical coordinates (r,ϕ,z) and the ray coordinates (σ,ϕ,a). Then we write $J = \left|\partial \underline{x} / \partial(r,\phi,z)\right| \left|\partial(r,\phi,z)/\partial(\sigma\ \phi,a)\right|$. The first factor is just r , and the second can be computed by using (6.17), (6.18) and (6.20). Thus we find

$$J = \frac{r}{n}(n^2-a^2)^{1/2} \int_{z_<}^{z_>} \frac{n^2}{(n^2-a^2)^{3/2}} dz \; . \tag{6.24}$$

We now use (6.24) and (6.15) in (6.14) to obtain

$$A_0(r,z) = \frac{a^{1/2}}{4\pi} [n^2(z_0)-a^2]^{-1/4} [n^2(z)-a^2]^{-1/4} \left[r\int_{z_<}^{z_>} \frac{n^2}{(n^2-a^2)^{3/2}} dz \right]^{-1/2} . \tag{6.25}$$

The leading term in the direct wave is $A_0(r,z)e^{ikS(r,z)}$ with A_0 given by (6.25) and S given by (6.23). This is exactly the same as the term with m=0 and i=j=1 in (5.84) for $p_{11}^{+}(r,z)$. To see this we first use (5.77) for $\partial^2 S_{11}^{+}/\partial a^2$ with m = 0 in (5.84). We then observe that (5.76) with m = 0 and j = 1 is the same as (6.20).

Therefore the root b_m of (5.76) is just the value of the ray parameter a for which the direct ray passes through (r,z) . Thus the identity of the two expressions is shown.

By proceeding in the same way, we can calculate the leading term in each singly and multiply reflected and refracted wave. Each one turns out to be identical with one of the terms in (5.84). Thus the total field p(r,z) given by the ray representation is exactly the same as (5.75) with the p_{ij}^+ given by (5.84).

References

[1] L. Brekhovskikh, Waves in Layered Media, Academic Press, New York, 1960.

[2] W. Magnus and F. Oberhettinger, Formulas and Theorems for the Functions of Mathematical Physics, Chelsea Publishing Company, New York, 1954.

[3] I. S. Gradshteyn and I. M. Ryzhik, Table of Integrals, Series and Products, Academic Press, New York, 1965.

[4] M. Abramowitz and I. A. Stegun, Handbook of Mathematical Functions, Nat. Bur. St., Appl. Math. Series, 55, 1964.

[5] P. M. Morse and H. Feshbach, Methods of Theoretical Physics, Vol. II, McGraw-Hill Book Company, Inc., New York, 1953.

[6] R. Lynn and J. B. Keller, Uniform asymptotic solutions of second order linear ordinary differential equations with turning points, Comm. on Pure and Appl. Math., 23, 1970, pp. 379-408.

[7] R. E. Langer, The asymptotic solutions of ordinary linear differential equations of the second order, with special reference to a turning point, Trans. Amer. Math. Soc., 67, 1949, pp. 461-490.

[8] A. Erdèlyi, Asymptotic Expansions, Dover Publications,Inc., New York, 1956.

[9] D. V. Batorsky and L. B. Felsen, Ray-optical calculation of modes exited by sources and scatterers in a weakly inhomogeneous duct, Radio Science, 6, 1971, pp. 911-923.

[10] D. S. Ahluwalia, R. M. Lewis and J. Boersma, Uniform asymptotic theory of diffraction by a plane screen, SIAM J. Appl. Math., 16, 1968, pp. 783-807.

[11] G. S. S. Avila and J. B. Keller, The high-frequency asymptotic field of a point source in an inhomogeneous medium, 16, 1963, pp. 363-381.

[12] W. M. Ewing, W. S. Jardetzky and F. Press, Elastic waves in layered media, McGraw-Hill Book Company, Inc., New York, 1957.

[13] L. B. Felsen and N. Marcuvitz, Radiation and Scattering of waves, Prentice Hall, New Jersey, 1973.

[14] J. B. Keller and R. M. Lewis, Asymptotic theory of wave propagation and diffraction, John Wiley, New York (to be published).

CHAPTER III

HORIZONTAL RAYS AND VERTICAL MODES

Robert Burridge

Courant Institute of Mathematical Sciences
New York University
251 Mercer Street
New York, NY 10012

Henry Weinberg

Naval Underwater Systems Center
New London Laboratory
New London, CT 06320

1. Introduction

In recent years, there has been a growing interest in long-range, low-frequency acoustic propagation in the ocean. Very often three-dimensional ray-tracing techniques are used to analyze this problem, but as the ranges increase and the frequencies of interest decrease, this kind of ray-tracing loses its effectiveness.

The theory of normal modes offers an alternative approach. Pekeris [1] was the first to apply the theory to underwater acoustics and his results for shallow water were later shown to agree well with the experimental data gathered by Worzel and Ewing [2] whose analysis was mainly concerned with dispersion. Tolstoy [3] later showed that wave-guide theory could be used to predict intensity levels in shallow water. However, the analyses of Pekeris and of Tolstoy require the medium to be perfectly stratified, that is, the properties of the medium are assumed constant on horizontal planes. Pierce [4] extended the theory to media having a slow variation in the horizontal directions under the assumption that the coupling between modes can be neglected. This assumption is borne out to zero order in the slowness of horizontal variation by our analysis. The principal results of his calculations were that different modes follow different horizontal paths and that the intensities along these horizontal rays satisfy transport equations in two space dimensions.

In this paper we shall also be concerned with acoustic propagation in an almost stratified medium which we shall take to represent the ocean. The method consists in introducing a small parameter ε representing the (slow) rate of variation of the medium in the horizontal directions. The velocity potential is sought in the form of an asymptotic power series in ε where the vertical structure is expressed in terms of the normal mode eigenfunctions. We find an eikonal equation and a recursive system of transport equations for the coefficients which depend only upon the horizontal coordinates. This scheme is closely analogous to geometrical acoustics in two dimensions. It was first described by Keller[5] and later used by Shen and Keller[6] in the context of surface waves on water of variable depth. A similar theory has also been developed by Rulf[7] and by Bretherton[8]. Our method has much in common with these but in the systematic use of the small parameter it is more in the spirit of the geometrical theory developed by Keller and his coworkers for scalar and vector wave equations[9,10]. The work reported here is a slight expansion of a paper by Weinberg and Burridge which appeared as reference[11].

In sections 2 and 3 we treat time-harmonic disturbances by considering solutions to the reduced wave equation. The small parameter ε is introduced in section 2 where we suppose that the properties of the medium depend upon the horizontal coordinates X, Y only through the combinations $x = \varepsilon X$, $y = \varepsilon Y$. This being so we seek at first a solution where the velocity potential ϕ is expressed in the form

$$1.1 \qquad \phi(x,y,z;\varepsilon) \sim e^{\theta(x,y)/(i\varepsilon)} \sum_{\nu=0}^{\infty} A_\nu(x,y,z)\,(i\varepsilon)^\nu ,$$

z being the vertical coordinate and a factor $e^{-\omega t/(i\varepsilon)}$ is understood. Each A_ν is expanded in the eigenfunctions $\psi_m(x,y;z)$ of a certain differential operator in z whose coefficients depend parametrically upon x,y:

$$1.2 \qquad A_\nu(x,y,z) = \sum_{k=0}^{\infty} a_\nu^k(x,y)\,\psi_k(x,y;z) .$$

A_0 is found to be a pure mode in that it is a multiple of a single eigenfunction ψ_p. An eikonal equation is found for θ and then a recursive system of transport equations are found for the $a_\nu^k(x,y)$. These equations are virtually identical to the corresponding equations for the ordinary geometrical wave theory in two dimensions. The leading term found in this way agrees with Pierce's solution.

As usual (1.1) is not valid near caustics. The necessary modifications are discussed in section 3. There we draw heavily from the work of Ludwig[12] and Babich [13] .

In section 4 a more general time dependence is considered in connection with the full wave equation. We use a generalization of the ansatz (1.1) in which the phase function and the coefficients may depend upon $t = \varepsilon T$ in addition to the space variables. Attention there is restricted to the leading coefficient A_0, which in practice yields a good approximation whenever the ray theory is valid. We consider the Airy phase, which is a space-time analog of the smooth caustic and suggest that some further work might be done in connection with the high frequency arrival or water wave and with modes propagating near cut-off. In this section at no further cost the wave speed is allowed to depend upon εT.

Two typical special cases are studied in section 5. The first concerns acoustic propagation in an ocean with constant sound speed but where the bottom depth varies linearly with Y. The ray configurations are computed and plotted for various propagating modes. Pierce[4] considered a similar model in which the reciprocal of the bottom depth varied linearly. In the second example the square of the wave number decreases linearly with depth.

Finally in section 6 a realistic model ocean is considered. The acoustic amplitudes are computed and found to agree well with real data. Also in section 6 is a brief description of the computer program which determines the normal modes, solves the ray equations, and finds the field quantities.

2. Acoustic propagation in an almost stratified medium

An almost stratified medium is a medium whose properties vary slowly with the horizontal coordinates. This notion is made precise by introducing a small parameter ε and supposing the wave number k depends upon X,Y, the horizontal coordinates, only through the combinations εX, εY. Moreover the boundaries will be taken as almost horizontal in the sense that on the boundaries the vertical coordinate Z will be given as a function of εX and εY.

Hence if $\phi e^{i\omega T}$ is the velocity potential in an almost stratified ocean then ϕ satisfies the reduced wave equation

2.1 $$\frac{\partial^2\phi}{\partial X^2} + \frac{\partial^2\phi}{\partial Y^2} + \frac{\partial^2\phi}{\partial Z^2} + k^2(\varepsilon X,\varepsilon Y,Z)\ \phi = 0$$

in the region

2.2 $$Z^-(\varepsilon X,\ \varepsilon Y) < Z < Z^+(\varepsilon X,\ \varepsilon Y)\ ,$$

where $k^2 = \omega^2/c^2$, $c(\varepsilon X,\varepsilon Y,Z)$ being the sound speed.

On the boundaries $Z = Z^{\pm}(\varepsilon X,\ \varepsilon Y)$ we assume the boundary conditions take the form

2.3 $$\alpha^{\pm}(\varepsilon X,\ \varepsilon Y)\ \phi \pm \beta^{\pm}(\varepsilon X,\ \varepsilon Y)\ \frac{\partial\phi}{\partial n^{\pm}} = 0\ ,$$

where $\alpha^{\pm}$, $\beta^{\pm}$ are real and $\partial/\partial n^{\pm}$ represents the outward normal derivative. In fact

2.4 $$\pm\ \frac{\partial\phi}{\partial n^{\pm}} = \frac{\dfrac{\partial\phi}{\partial Z} - \dfrac{\partial Z^{\pm}}{\partial X}\dfrac{\partial\phi}{\partial X} - \dfrac{\partial Z^{\pm}}{\partial Y}\dfrac{\partial\phi}{\partial Y}}{\left[1 + \left(\dfrac{\partial Z^{\pm}}{\partial X}\right)^2 + \left(\dfrac{\partial Z^{\pm}}{\partial Y}\right)^2\right]^{1/2}}\ .$$

Upon introducing new coordinates in which the horizontal distances are contracted,

2.5a $$x = \varepsilon\ X\ ,$$

2.5b $$y = \varepsilon\ Y\ ,$$

2.5c $$z = Z\ ,$$

into the above equations we obtain

$$\nabla^2\phi - (i\varepsilon)^{-2} L\,\phi = 0 \tag{2.6}$$

with boundary conditions

$$\alpha^{\pm}(x,y)\phi + \beta^{\pm}(x,y)\gamma^{\pm}(x,y;\varepsilon)\left(\frac{\partial\phi}{\partial z} + (i\varepsilon)^2\nabla Z^{\pm}\cdot\nabla\phi\right) = 0 \tag{2.7}$$

at $z = Z^{\pm}(x,y)$.

In (2.6) and (2.7) ∇ is the horizontal gradient operator $(\partial/\partial x\ ,\ \partial/\partial y)$, L is the operator given by

$$L\,\phi \equiv \frac{\partial^2\phi}{\partial z^2} + k^2(x,y,z)\,\phi\ , \tag{2.8}$$

and $\gamma^{\pm}$ is given by

$$\gamma^{\pm}(x,y;\varepsilon) = \frac{1}{\left[1 - (i\varepsilon)^2\ (\nabla Z^{\pm})^2\right]^{1/2}}\ . \tag{2.9}$$

By analogy with existing geometrical wave theory we shall seek ϕ in the form of a series

$$\phi(x,y,z;\varepsilon) \sim e^{\theta(x,y)/(i\varepsilon)} \sum_{\nu=0}^{\infty} (i\varepsilon)^{\nu}\ A_{\nu}(x,y,z)\ , \tag{2.10}$$

which is interpreted to mean that

$$\sum_{\nu=0}^{\infty} (i\varepsilon)^{\nu}\ A_{\nu}\ (x,y,z) \text{ is asymptotic to } e^{-\theta/i\varepsilon}\ \phi(x,y,z;\varepsilon)$$

as $\varepsilon \to 0$.

Substitution of (2.10) into (2.6) and boundary conditions (2.7) will lead in the usual way to an eikonal equation for θ and to various transport equations. In order to perform the substitution we shall need

$$2.11 \qquad \nabla \phi = e^{\frac{\theta}{i\varepsilon}} \sum_{\nu=0}^{\infty} (\nabla\theta A_{\nu} + \nabla A_{\nu-1}) (i\varepsilon)^{\nu-1}$$

and

$$2.12 \qquad \nabla^2 \phi = e^{\frac{\theta}{i\varepsilon}} \sum_{\nu=0}^{\infty} \left[(\nabla\theta)^2 A_{\nu} + \nabla^2\theta A_{\nu-1} + 2\, \nabla\theta\cdot\nabla A_{\nu-1} + \nabla^2 A_{\nu-2} \right] (i\varepsilon)^{\nu-2} ,$$

where for convenience in notation $A_{-1} \equiv 0$, $A_{-2} \equiv 0$.
Substitution into (2.6) now gives

$$2.13 \qquad \sum_{\nu=0}^{\infty} \left[(\nabla\theta)^2 A_{\nu} - LA_{\nu} + \nabla^2\theta A_{\nu-1} + 2\, \nabla\theta\cdot\nabla A_{\nu-1} + \nabla^2 A_{\nu-2} \right] (i\varepsilon)^{\nu-2} = 0 .$$

Before substituting in the boundary conditions let us write $\gamma^{\pm}$ as power series in $i\varepsilon$:

$$2.14 \qquad \gamma^{\pm}(x,y;\varepsilon) = \Sigma\, \gamma_q^{\pm}\, (x,y)\, (i\varepsilon)^{2q} ,$$

where

$$2.15 \qquad \gamma_q^{\pm} = (-1)^q \binom{-1/2}{q} (\nabla Z^{\pm})^{2q} .$$

Hence from 2.7

$$2.16 \qquad \alpha^{\pm}\phi + \beta^{\pm} \sum_{q=0}^{\infty} \gamma_q^{\pm} \left[\frac{\partial\phi}{\partial z} + (i\varepsilon)^2\, \nabla Z^{\pm} \cdot \nabla\phi \right] (i\varepsilon)^{2q} = 0 .$$

On substituting (2.10) into (2.16) and using (2.11) we have, after some rearrangement, that on $z = Z^{\pm}(x,y)$

$$\alpha^{\pm} \sum_{\nu=0}^{\infty} (i\varepsilon)^{\nu} A_{\nu} + \beta^{\pm} \sum_{\nu=0}^{\infty} \left\{ \sum_{q=0}^{\nu_1} \gamma_q^{\pm} \left[\frac{\partial A_{\nu-2q}}{\partial z} + \nabla Z^{\pm} \cdot \nabla\theta A_{\nu-2q-1} + \nabla Z^{\pm} \cdot \nabla A_{\nu-2q-2} \right] (i\varepsilon)^{\nu} \right\} = 0 \;, \tag{2.17}$$

where $\nu_1 = [\nu/2]$, is the largest integer not greater than $\nu/2$.

The assumption of the asymptotic nature of (2.10) allows us to equate to zero the coefficients of individual powers of $i\varepsilon$ in (2.13) and (2.17) to get

$$(\nabla\theta)^2 A_{\nu} - LA_{\nu} + \nabla^2\theta A_{\nu-1} + 2\nabla\theta \cdot \nabla A_{\nu-1} + \nabla^2 A_{\nu-2} = 0 \;, \qquad \nu = 0,1,2,\ldots, \tag{2.18}$$

and on $z = Z^{\pm}$,

$$\alpha^{\pm} A_{\nu} + \beta^{\pm} \sum_{q=0}^{\nu_1} \gamma_q^{\pm} \left[\frac{\partial A_{\nu-2q}}{\partial z} + \nabla Z^{\pm} \cdot \nabla\theta A_{\nu-2q-1} + \nabla Z^{\pm} \cdot \nabla A_{\nu-2q-2} \right] = 0 \;. \tag{2.19}$$

In (2.19) let us write the terms in A_{ν} on the left and separate the terms in $A_{\nu-1}$ from the rest of the right side:

$$\alpha^{\pm} A_{\nu} + \beta^{\pm} \frac{\partial A_{\nu}}{\partial z} = -\beta^{\pm}[\nabla Z^{\pm} \cdot \nabla\theta A_{\nu-1} + F_{\nu-2}^{\pm}(x,y;A_0,\ldots,A_{\nu-2})], \; \nu = 0,1,2,\ldots, \tag{2.20}$$

where

$$F_{\nu-2}^{\pm}(x,y;A_0,\ldots,A_{\nu-2}) \equiv \nabla Z^{\pm} \cdot \nabla A_{\nu-2} + \sum_{q=1}^{\nu_1} \gamma_q^{\pm} \left[\frac{\partial A_{\nu-2q}}{\partial z} + \nabla Z^{\pm} \cdot \nabla\theta A_{\nu-2q-1} + \nabla Z^{\pm} \cdot \nabla A_{\nu-2q-2} \right] . \tag{2.21}$$

Upon setting $\nu=0$ in (2.18) and (2.20) we obtain

$$(\nabla\theta)^2 A_0 - LA_0 = 0 \tag{2.22}$$

and

2.23 $$\alpha^{\pm} A_o + \beta^{\pm} \frac{\partial A_o}{\partial z} = 0 .$$

Equations (2.22) and (2.23) define an eigenvalue problem which shows that $(\nabla\theta)^2$ is an eigenvalue of L and A_o a corresponding eigenfunction.

Let

2.24 $\lambda_o^2 , \lambda_1^2 , \lambda_2^2 , \ldots$ be the eigenvalues of L .

We shall suppose that all eigenvalues are simple. Let

2.25 $\psi_o , \psi_1 , \psi_2 , \ldots$ be the corresponding eigenfunctions

normalized with respect to the inner product of (2.30).

Suppose that

2.26 $$(\nabla\theta)^2 = \lambda_p^2(x,y) .$$

This is the eikonal equation for θ and may be solved for θ by the method of characteristics. In fact (2.26) leads to the ray equations

2.27a $$\frac{d}{ds}\left(\lambda_p \frac{dx}{ds}\right) = \frac{\partial \lambda_p}{\partial x} ,$$

2.27b $$\frac{d}{ds}\left(\lambda_p \frac{dy}{ds}\right) = \frac{\partial \lambda_p}{\partial y} ,$$

2.28 $$\frac{d\theta}{ds} = \lambda_p ,$$

where

2.29 $$\left(\frac{dx}{ds}\right)^2 + \left(\frac{dy}{ds}\right)^2 = 1 .$$

The rays are the curves satisfying (2.27a) and (2.27b) and the arc length is denoted by s. Once initial values of $\frac{dx}{ds}$, $\frac{dy}{ds}$, θ are given at a point, (2.27a) and (2.27b) can be integrated to obtain the ray whose initial direction is $(\frac{dx}{ds}, \frac{dy}{ds})$ and (2.28) can be (simultaneously) integrated to find θ at points along the ray. We must, of course, already know λ_p as a function of x,y. Notice that since L, $\alpha^{\pm}$, $\beta^{\pm}$ depend continuously on x, y, so do the λ_i.

Before proceeding to find the equations for the A_ν let us define an inner product on the space of functions of z, $Z^-(x,y) < z < Z^+(x,y)$.
We set

2.30
$$\langle f,g\rangle = \langle f(x,y;z), g(x,y;z)\rangle \equiv \int_{Z^-(x,y)}^{Z^+(x,y)} f(x,y;z)g(x,y;z)\, dz .$$

It can readily be shown that the ψ_m of (2.25) can be normalized to satisfy

2.31
$$\langle \psi_m, \psi_n\rangle = \delta_{mn} ,$$

where δ_{mn} is the Kronecker delta.

From (2.22), (2.25) and (2.26) it is clear that

2.32
$$A_o(x,y,z) = a_o(x,y)\, \psi_p(x,y;z) .$$

We shall now obtain a transport equation for a_o. Set $\nu = 1$ in (2.18) and (2.20) to get

2.33
$$\lambda_p^2 A_1 - LA_1 + \nabla^2\theta A_o + 2\nabla\theta\cdot\nabla A_o = 0 ,$$

and

2.34
$$\alpha^{\pm} A_1 + \beta^{\pm} \frac{\partial A_1}{\partial z} = -\beta^{\pm} \nabla Z^{\pm}\cdot\nabla\theta\, A_o$$

on $z = Z^{\pm}(x,y)$.

Now take the inner product of (2.33) with ψ_p :

2.35 $$\lambda_p^2 \langle\psi_p, A_1\rangle - \langle\psi_p, LA_1\rangle + \nabla^2\theta\ a_o + 2\nabla\theta\cdot\langle\psi_p, \nabla A_o\rangle = 0\ .$$

But on using (2.23) and (2.34) we find that

2.36 $$\langle\psi_p, LA_1\rangle = \lambda_p^2\langle\psi_p, A_1\rangle - a_o\ \nabla\theta\cdot[\psi_p^2 \nabla Z]_-^+\ ,$$

where

2.37 $$[\psi_p^2\ \nabla Z]_-^+ \equiv \psi_p^2\ (x,y;Z^+)\ \nabla Z^+(x,y) - \psi_p^2\ (x,y;Z^-)\nabla Z^-(x,y)\ .$$

Also since $\langle\psi_p, \psi_p\rangle = 1$ it follows on taking gradients that

2.38 $$2\langle\psi_p, \nabla\psi_p\rangle = -[\psi_p^2\ \nabla Z]_-^+\ .$$

Thus (2.35) reduces to

2.39 $$\nabla^2\theta\ a_o + 2\nabla\theta\cdot\nabla a_o = 0\ ,$$

which is the transport equation for a_o .

Equation (2.39) may also be written as

2.40 $$\nabla\cdot(a_o^2\ \nabla\theta) = 0\ .$$

On integrating (2.40) over a ray tube and using the divergence theorem we obtain

2.41 $\lambda_p a_o^2\ \delta\sigma$ is constant along the tube,

where $\delta\sigma$ is the cross-sectional width of the tube of rays.

From the Sturm-Liouville theory of self-adjoint ordinary differential equations we know that the functions ψ_m, m=0,1,2, form a complete set. Therefore we shall assume that

2.42 $$A_\nu(x,y;z) = \sum_{k=0}^{\infty} a_\nu^k\,(x,y)\;\psi_k(x,y;z) + G_\nu(x,y;z)\ ,$$

where G_ν is any function satisfying the boundary conditions

2.43 $$\alpha^{\pm} G_\nu + \beta^{\pm}\frac{\partial G_\nu}{\partial z} = -\beta^{\pm}[\nabla Z\cdot\nabla\theta A_{\nu-1} + F^{\pm}_{\nu-2}(x,y;A_o,\ldots,A_{\nu-2})]$$

on $z = Z^{\pm}(x,y)$.

Suppose, now, that the a_μ^k are known for all k and $\mu = 0, 1, 2, \ldots, \nu - 1$. Then we may determine G_ν to satisfy (2.43). On substituting (2.42) into (2.18) and taking the scalar product with ψ_m , $m \neq p$, we get

2.44 $$a_\nu^m = (\lambda_m^2 - \lambda_p^2)^{-1} <\psi_m, \nabla^2\theta A_{\nu-1} + 2\nabla\theta\cdot\nabla A_{\nu-1} + \nabla^2\,A_{\nu-2} + (\lambda_p^2 - L)G_\nu>.$$

To find a_ν^p we take (2.18) with ν replaced by $\nu + 1$ and take the inner product with ψ_p to get

2.45 $$\lambda_p^2 <\psi_p, A_{\nu+1}> - <\psi_p, LA_{\nu+1}>$$
$$+ \nabla^2\theta<\psi_p,\ A_\nu> + 2\nabla\theta\cdot<\psi_p,\ \nabla A_\nu> + <\psi_p,\ \nabla^2 A_{\nu-1}> = 0\ .$$

But

2.46 $$<\psi_p,\ LA_{\nu+1}> = <L\psi_p,\ A_{\nu+1}> + \left[\psi_p\,\frac{\partial A_{\nu+1}}{\partial z} - \frac{\partial\psi_p}{\partial z}\,A_{\nu+1}\right]_-^+$$
$$= \lambda_p^2<\psi_p, A_{\nu+1}> + \left[\psi_p\left(\frac{\partial A_{\nu+1}}{\partial z} + \frac{\alpha}{\beta}\,A_{\nu+1}\right)\right]_-^+$$
$$= \lambda_p^2<\psi_p, A_{\nu+1}> + \left[\psi_p\left(-\nabla Z\cdot\nabla\theta A_\nu - F_{\nu-1}\right)\right]_-^+$$

so that from (2.45)

2.47 $$\nabla^2\theta \langle\psi_p, A_\nu\rangle + 2\,\nabla\theta\cdot\langle\psi_p,\nabla A_\nu\rangle + \langle\psi_p, \nabla^2 A_{\nu-1}\rangle$$

$$= - \Big[\psi_p \nabla Z\cdot\nabla\theta A_\nu\Big]_-^+ - \Big[\psi_p\, F_{\nu-1}\Big]_-^+ \quad .$$

Substituting (2.42) into (2.47) gives

2.48 $$\nabla^2\theta\, a_\nu^p + 2\,\nabla\theta\cdot\nabla a_\nu^p = - \nabla^2\theta\cdot\langle\psi_p, G_\nu\rangle - 2\nabla\theta\cdot\langle\psi_p, \nabla G_\nu\rangle$$

$$- 2\nabla\theta\cdot \sum_{k=0}^{\infty} a_\nu^k \langle\psi_p, \nabla\psi_k\rangle - \langle\psi_p, \nabla^2 A_{\nu-1}\rangle$$

$$- \left[\psi_p \nabla Z\cdot\nabla\theta\left(\sum_{k=0}^{\infty} a_\nu^k\,\psi_k + G_\nu\right)\right]_-^+ - \Big[\psi_p\, F_{\nu-1}\Big]_-^+ \quad .$$

We now notice that the $k = p$ terms in the two sums in (2.48) cancel so that the right member of (2.48) involves only functions already determined, including the a_ν^k for $k \neq p$. As usual the left member is a directional derivative of a_ν^p along a ray.

Thus (2.44) and (2.48) form a system from which the a_ν^k can be found recursively.

3. Uniform asymptotic expansions in regions containing caustics

We shall be concerned in sections 5 and 6 with the field due to a point source. The horizontal rays all pass through the source point and moreover in the examples treated these rays envelop curves called caustics. Whenever neighboring rays come together as at a point source or at a caustic the simple theory of section 2 becomes invalid in the neighborhood of the points of concurrence. In this section we shall show how the ray theory may be modified to treat first the field near a smooth caustic and then we shall discuss certain difficulties which prevent us from satisfactorily dealing with the point source.

3.1 The field near a smooth caustic

It will be seen in section 5 that in quite simple cases rays emanating from a point source will envelop a caustic. When this occurs neighboring rays come together and the theory of section 2 would predict an infinite amplitude. To obtain a valid approximation near a caustic we shall follow Ludwig [12] and seek an asymptotic expression for ϕ in the form

3.1 $$\phi(x,y,z,\varepsilon) \sim e^{\frac{\theta}{i\varepsilon}} \sum_{\nu=0}^{\infty} \Big[(i\varepsilon)^{\nu} A_{\nu}(x,y,z)\, V[\varepsilon^{-2/3}\rho(x,y)] + (i\varepsilon)^{\nu}(i\varepsilon^{1/3})\, B_{\nu}(x,y,z)\, V'[\varepsilon^{-2/3}\rho(x,y)] \Big],$$

where $V(\zeta)$ is the Airy function of $-\zeta$ and so satisfies

3.2 $$V''(\zeta) + \zeta V(\zeta) = 0 .$$

Substitution of the expansion (3.1) into (2.6) yields

3.3 $$\begin{aligned} e^{\frac{\theta}{i\varepsilon}} \sum_{\nu=0}^{\infty} \Big[& (i\varepsilon)^{\nu-2} \{[(\nabla\theta)^2 + \rho(\nabla\rho)^2 - L]A_{\nu} + 2\rho\nabla\rho\cdot\nabla\theta B_{\nu}\} V \\ & + (i\varepsilon)^{\nu-2}\{[(\nabla\theta)^2 + \rho(\nabla\rho)^2 - L]B_{\nu} + 2\nabla\rho\cdot\nabla\theta A_{\nu}\}(i\varepsilon^{1/3}) V' \\ & + (i\varepsilon)^{\nu-1}\{\nabla^2\theta A_{\nu} + 2\nabla\theta\cdot\nabla A_{\nu} + \rho\nabla^2\rho\, B_{\nu} + 2\rho\nabla\rho\cdot\nabla B_{\nu} + (\nabla\rho)^2 B_{\nu}\} V \\ & + (i\varepsilon)^{\nu-1}\{\nabla^2\theta B_{\nu} + 2\nabla\theta\cdot\nabla B_{\nu} + \nabla^2\rho A_{\nu} + 2\nabla\rho\cdot\nabla A_{\nu}\}(i\varepsilon^{1/3}) V' \\ & + (i\varepsilon)^{\nu}\nabla^2 A_{\nu} V + (i\varepsilon)^{\nu} B_{\nu}(i\varepsilon^{1/3}) V' \Big] = 0 . \end{aligned}$$

The boundary conditions (2.16) lead to

3.4 $$\begin{aligned} e^{\frac{\theta}{i\varepsilon}} \sum_{\nu=0}^{\infty} (i\varepsilon)^{\nu} \Big\{ & \alpha(A_{\nu} V + i\varepsilon^{1/3} B_{\nu} V') \\ & + \beta \sum_{q=0}^{\nu_1} \gamma_q \Big[\Big(\frac{\partial A_{\nu-2q}}{\partial z} V + i\varepsilon^{1/3} \frac{\partial B_{\nu-2q}}{\partial z} V' \Big) \\ & + \Big(\rho\nabla\rho\cdot\nabla Z\, B_{\nu-2q-1} + \nabla\theta\cdot\nabla Z A_{\nu-2q-1} + \nabla Z\cdot\nabla A_{\nu-2q-2} \Big) V \\ & + i\varepsilon^{1/3} \Big(\nabla\rho\cdot\nabla Z A_{\nu-2q-1} + \nabla\theta\cdot\nabla Z B_{\nu-2q-1} + \nabla Z\cdot\nabla B_{\nu-2q-2} \Big) V' \Big] \Big\} = 0 , \end{aligned}$$

on $z = Z(x,y)$ where for clarity we have dropped the superscript $\pm$, and $\nu_1 = [\nu/2]$ as usual.

Equating coefficients of $(i\varepsilon)^\nu V$ and $(i\varepsilon)^\nu (i\varepsilon^{1/3}) V'$ in (3.3) separately to zero we obtain

3.5a
$$[(\nabla\theta)^2 + \rho(\nabla\rho)^2 - L]\, A_\nu + 2\rho\nabla\rho\cdot\nabla\theta B_\nu$$
$$+ \nabla^2\theta A_{\nu-1} + 2\nabla\theta\cdot\nabla A_{\nu-1} + \rho\nabla^2\rho B_{\nu-1} + 2\rho\nabla\rho\cdot\nabla B_{\nu-1} + (\nabla\rho)^2 B_{\nu-1}$$
$$+ \nabla^2 A_{\nu-2} = 0 ,$$

and

3.5b
$$[(\nabla\theta)^2 + \rho(\nabla\rho)^2 - L]\, B_\nu + 2\nabla\rho\cdot\nabla\theta A_\nu$$
$$+ \nabla^2\theta B_{\nu-1} + 2\nabla\theta\cdot\nabla B_{\nu-1} + \nabla^2\rho A_{\nu-1} + 2\nabla\rho\cdot\nabla A_{\nu-1}$$
$$+ \nabla^2 B_{\nu-2} = 0 .$$

Similarly (3.4) gives

3.6a
$$\alpha A_\nu + \beta \sum_{q=0}^{\nu_1} \gamma_q \left[\frac{\partial A_{\nu-2q}}{\partial z} + \rho\nabla\rho\cdot\nabla Z B_{\nu-2q-1} + \nabla\theta\cdot\nabla Z A_{\nu-2q-1} + \nabla Z\cdot\nabla A_{\nu-2q-2}\right] = 0 ,$$

and

3.6b
$$\alpha B_\nu + \beta \sum_{q=0}^{\nu_1} \gamma_q \left[\frac{\partial B_{\nu-2q}}{\partial z} + \nabla\rho\cdot\nabla Z A_{\nu-2q-1} + \nabla\theta\cdot\nabla Z B_{\nu-2q-1} + \nabla Z\cdot\nabla B_{\nu-2q-2}\right] = 0 .$$

Once again separating terms in A_ν , B_ν from the rest we get

3.7a $$\alpha A_\nu + \beta \frac{\partial A_\nu}{\partial z} = - \beta[\rho\nabla\rho\cdot\nabla ZB_{\nu-1} + \nabla\theta\cdot\nabla ZA_{\nu-1} + E_{\nu-2}]$$

3.7b $$\alpha B_\nu + \beta \frac{\partial B_\nu}{\partial z} = - \beta[\nabla\rho\cdot\nabla ZA_{\nu-1} + \nabla\theta\cdot\nabla ZB_{\nu-1} + F_{\nu-2}]$$

where $E_{\nu-2}$, $F_{\nu-2}$ depend on A_μ , B_μ , $0 \leq \mu \leq \nu - 2$ but not $A_{\nu-1}$, A_ν , $B_{\nu-1}$, B_ν . Setting $\nu = 0$ in (3.5a), (3.5b), (3.7a), and (3.7b) we see that, if we require that

3.8 $$\nabla\theta\cdot\nabla\rho = 0 ,$$

then

3.9 $$(\nabla\theta)^2 + \rho(\nabla\rho)^2 = \lambda_p^2 ,$$

where λ_p^2 is an eigenvalue of L and A_o , B_o are eigenfunctions of L with the boundary conditions

3.10a $$\alpha A_o + \beta \frac{\partial A_o}{\partial z} = 0 ,$$

3.10b $$\alpha B_o + \beta \frac{\partial B_o}{\partial z} = 0 .$$

Therefore we set

3.11a $$A_0(x,y,z) = a_0(x,y)\ \psi_p(x,y;z) ,$$

3.11b $$B_0(x,y,z) = b_0(x,y)\ \psi_p(x,y;z) .$$

On setting $\nu = 1$ in (3.5a) and(3.5b) and taking inner products with ψ_p we obtain

3.12a $$2\nabla\theta\cdot\nabla a_o + \nabla^2\theta a_o + 2\rho\nabla\rho\cdot\nabla b_o + \rho\nabla^2\rho b_o + (\nabla\rho)^2 b_o = 0$$

and

$$2\nabla\theta\cdot\nabla b_o + \nabla^2\theta b_o + 2\nabla\rho\cdot\nabla a_o + \nabla^2\rho a_o = 0 \; . \tag{3.12b}$$

As in section 2 the boundary terms have canceled by virtue of (3.7a) and (3.7b) with $\nu = 1$.

We note that (3.8), (3.9), (3.12a), and (3.12b) reduce, as they should, to the corresponding equations given by Ludwig [11] when we set $\lambda_p^2 = 1$. Following Ludwig let us write

$$\theta^{\pm} = \theta \pm 2/3\, \rho^{3/2} \; , \tag{3.13}$$

and

$$a_o^{\pm} = a_o \pm \rho^{1/2}\, b_o \; . \tag{3.14}$$

(Here $\pm$ refers to whether or not the ray has touched the caustic and has no connection with the superscripts referring to the boundaries $Z^{\pm}$.) Then it follows from (3.8) and (3.9) that

$$(\nabla\theta^{\pm})^2 = \lambda_p^2 \; . \tag{3.15}$$

By forming the linear combination $(3.12a)\pm\rho^{1/2}\times(3.12b)$ we also have

$$2\nabla\theta^{\pm}\cdot\nabla a_o^{\pm} + [\nabla^2\theta^{\pm} \mp 1/2\, \rho^{-1/2}(\nabla\rho)^2]\, a_o^{\pm} = 0 \; , \tag{3.16}$$

which reduces to

$$2\nabla\theta^{\pm}\cdot\nabla\, (\rho^{-1/4}\, a_o^{\pm}) + \nabla^2\theta^{\pm}\, (\rho^{-1/4}\, a_o^{\pm}) = 0 \; . \tag{3.17}$$

Thus $\theta^{\pm}$ may be found by solving the ordinary eikonal equation. The combinations $\rho^{-1/4}\, a_o^{\pm}$ are seen to satisfy (3.17), the ordinary zero order transport equation.

At the caustic ρ becomes zero and $\rho^{-1/4}\ a_o^{\pm}$ become infinite in such a way that $a_o^{\pm}$ remain finite.

For A_ν, B_ν with $\nu > 0$ we set

3.18a $$A_\nu = \sum_{k=0}^{\infty} a_\nu^k \psi_k + G_\nu ,$$

3.18b $$B_\nu = \sum_{k=0}^{\infty} b_\nu^k \psi_k + H_\nu ,$$

3.19a $$\alpha G_\nu + \beta \frac{\partial G_\nu}{\partial z} = - \beta[\rho\nabla\rho\cdot\nabla Z B_{\nu-1} + \nabla\theta\cdot\nabla Z A_{\nu-1} + E_{\nu-2}] ,$$

3.19b $$\alpha H_\nu + \beta \frac{\partial H_\nu}{\partial z} = - \beta[\nabla\rho\cdot\nabla Z A_{\nu-1} + \nabla\theta\cdot\nabla Z B_{\nu-1} + F_{\nu-2}] .$$

Now suppose a_μ^k, b_μ^k are known for all $\mu = 1, 2, \ldots, \nu - 1$ and all k. Then G_ν and H_ν may be chosen to satisfy (3.19a) and (3.19b). For a_ν^k, b_ν^k, $k \neq p$ we use (3.5a) and (3.5b)

3.20a $$a_\nu^m = (\lambda_m^2 - \lambda_p^2)^{-1} \langle\psi_m, \nabla^2\theta A_{\nu-1} + 2\nabla\theta\cdot\nabla A_{\nu-1} + \rho\nabla^2\rho B_{\nu-1} + 2\rho\nabla\rho\cdot\nabla B_{\nu-1} + (\nabla\rho)^2 B_{\nu-1} + \nabla^2 A_{\nu-2} - (\lambda_p^2 - L)\, G_\nu\rangle ,$$

3.20b $$b_\nu^m = (\lambda_m^2 - \lambda_p^2)^{-1} \langle\psi_m, \nabla^2\theta B_{\nu-1} + 2\nabla\theta\cdot\nabla B_{\nu-1} + \nabla^2\rho A_{\nu-1} + 2\nabla\rho\cdot\nabla A_{\nu-1} + \nabla^2 B_{\nu-2} - (\lambda_p^2 - L)\, H_\nu\rangle .$$

For a_ν^p, b_ν^p we replace ν by $\nu + 1$ in (3.5a) and (3.5b) and take inner products with ψ_p. It is convenient to treat the terms in (3.5a) separately:

3.21 $$\langle\psi_p,[(\nabla\theta)^2 + \rho(\nabla\rho)^2 - L]\, A_{\nu+1}\rangle = \langle\psi_p,(\lambda_p^2 - L)\, A_{\nu+1}\rangle$$

$$= \frac{-1}{\beta}\,[\psi_p(\alpha A_{\nu+1} + \beta\,\frac{\partial A_{\nu+1}}{\partial z})]_-^+$$

$$= [\psi_p(\rho\nabla\rho\cdot\nabla Z B_\nu + \nabla\theta\cdot\nabla Z A_\nu) + \psi_p E_{\nu-1}]_-^+$$

$$= \Bigg[\rho\nabla\rho\cdot\nabla Z\psi_p \sum_{k=0}^{\infty} b_\nu^k\,\psi_k + H_\nu +$$

$$+ \nabla\theta\cdot\nabla Z\psi_p\left(\sum_{k=0}^{\infty} a_\nu^k\,\psi_k + G_\nu\right) + \psi_p E_{\nu-1}\Bigg]_-^+$$

by (3.7a). Also

3.22 $$\langle\psi_p,2\nabla\theta\cdot\nabla A_{\nu-1}\rangle = 2\nabla\theta\cdot\nabla a_\nu^p + 2\nabla\theta\cdot\langle\psi_p,\nabla G_\nu\rangle$$

$$+ 2\nabla\theta\cdot\sum_{k=0}^{\infty} a_\nu^k\langle\psi_p,\nabla\psi_k\rangle$$

$$= 2\nabla\theta\cdot\nabla a_\nu^p + 2\nabla\theta\cdot\langle\psi_p,\nabla G_\nu\rangle$$

$$+ 2\nabla\theta\cdot\sum_{k\neq p} a_\nu^k\langle\psi_p,\nabla\psi_k\rangle$$

$$- [\nabla\theta\cdot\nabla Z a_\nu^p\,\psi_p^2]_-^+ ,$$

where we have used (2.38). It will be noticed that the last term in (3.22) precisely cancels the term k = p in the second sum of (3.21) so that a_ν^p disappears from these contributions.

Similarly

3.23 $$\langle\psi_p,2\rho\nabla\rho\cdot\nabla B_\nu\rangle = 2\rho\nabla\rho\cdot\nabla b_\nu^p + 2\rho\nabla\rho\cdot\langle\psi_p,\nabla H_\nu\rangle$$

$$+ 2\rho\nabla\rho\cdot\sum_{k=0}^{\infty} b_\nu^k\langle\psi_p,\nabla\psi_k\rangle$$

$$= 2\rho\nabla\rho\cdot\nabla b_\nu^p + 2\rho\nabla\rho\cdot\langle\psi_p,\nabla H_\nu\rangle$$

$$+ 2\rho\nabla\rho\cdot \sum_{k\neq p} b_\nu^k\langle\psi_p,\nabla\psi_k\rangle$$

$$- [\rho\nabla\rho - \nabla Z b_\nu^p \psi_p^2]_-^+ ,$$

where we have used (2.38) once again. As before the last term cancels the term $\nu = p$ in the first sum of (3.21). Thus no term in b_ν^p arises from these contributions.

There remains from (3.5a) the following:

3.24 $$\langle\psi_p,\nabla^2\theta A_\nu\rangle + [\rho\nabla^2\rho + (\nabla\rho)^2]\langle\psi_p,B_\nu\rangle + \langle\psi_p,\nabla^2 A_{\nu-1}\rangle$$

$$= \nabla^2\theta a_\nu^p + [\rho\nabla^2\rho + (\nabla\rho^2)]\, b_\nu^p + \nabla^2\theta\langle\psi_p,G_\nu\rangle$$

$$+ [\rho\nabla^2 p + (\nabla\rho)^2]\,\langle\psi_p,H_\nu\rangle + \langle\psi_p,\nabla^2 A_{\nu-1}\rangle .$$

Collecting (3.21)-(3.24) together we have

3.25a $$2\nabla\theta\cdot\nabla a_\nu^p + \nabla^2\theta a_\nu^p + 2\rho\nabla\rho\cdot\nabla b_\nu^p + [\rho\nabla^2\rho + (\nabla\rho)^2]\, b_\nu^p$$

= a function of previously determined quantities.

Similarly from (3.5b) we may deduce that

3.25b $$2\nabla\theta\cdot\nabla b_\nu^p + \nabla^2\theta b_\nu^p + 2\nabla\rho\cdot\nabla a_\nu^p + \nabla^2\rho a_\nu^p$$

= a function of previously determined quantities.

(3.25a) and (3.25b) are the higher-order transport equations which as usual are inhomogeneous versions of the zero-order equations (3.12a) and (3.12b). The coefficients a_ν^m for $m \neq p$ are determined by the <u>algebraic</u> equations (3.20a) and (3.20b)

This completes our theory of the field near a smooth caustic.

3.2 A point source in an almost stratified medium

Whenever neighboring rays come together the theory of section 2 breaks down since (2.41) predicts infinite amplitudes. This situation arises in our applications not only at smooth caustics but in the neighborhood of a point source.

By analogy with the exact solution for a perfectly stratified medium it might be thought that an asymptotic solution of the form

$$3.26 \qquad \phi(x,y,z;\varepsilon) = \sum_{\nu=0}^{\infty} \{\varepsilon^{\nu} A_{\nu}(x,y,z)\, H_0^{(2)}[\varepsilon^{-1}\,\theta(x,y)] + \varepsilon^{\nu} B_{\nu}(x,y,z)\, H_0^{(2)\prime}[\varepsilon^{-1}\,\theta(x,y)]\} ,$$

where $H_0^{(2)}$ is the zeroth order Hankel function of the second kind, would be uniform near $\theta = 0$. Indeed, when the dependence upon z is absent this ansatz is valid for the scalar Helmholtz equation near a point source [14], and if the coefficients of the original equation are regular near $\theta = 0$ an argument of Hadamard [15] in his construction of the elementary solution shows that A_ν , B_ν are also regular.

A slightly neater alternative to (3.26) due to Babich [13] is also uniform near the source for the scalar (z-independent) equation. This is

$$3.27 \qquad \phi(x,y,z;\varepsilon) = \sum_{\nu=0}^{\infty} A_{\nu}(x,y,z)\, f_{\nu}(\varepsilon,\theta) ,$$

where

$$3.28 \qquad f_{\nu}(\varepsilon,\theta) = \varepsilon^{\nu}\, \theta^{\nu}\, H_{\nu}^{(2)}\, (\theta/\varepsilon) .$$

However, neither (3.26) nor (3.27) are valid for our problem unless, at $\theta = 0$, $\langle\psi_m , \nabla\psi_n\rangle = 0$. This can arise, for instance, when the ocean has axial symmetry about the vertical through the source point -- an unpleasantly restrictive hypothesis.

Let us illustrate the difficulty which arises on using (3.27) and (3.28). We first note the recurrence formulae for the f_ν:

3.29a $$\frac{1}{\varepsilon^2} f_\nu = 2(\nu - 1) f_{\nu-1} - \theta^2 f_{\nu-2}$$

3.29b $$\frac{\partial f_\nu}{\partial \theta} = \theta f_{\nu-1} ,$$

On substituting (3.27) into (2.6) and using (3.29a) and (3.29b) we obtain

3.30 $$\sum_{\nu=0}^{\infty} \{\theta^2[(\nabla\theta)^2 A_\nu - LA_\nu] + [\nabla(\theta^2)\cdot\nabla A_{\nu-1} + 1/2\ \nabla^2(\theta^2)A_{\nu-1} + 2(\nu-2)LA_{\nu-1}] + \nabla^2 A_{\nu-2}\}\ f_{\nu-2} = 0 .$$

On equating the coefficient of f_{-2} to zero we get

3.31 $$(\nabla\theta)^2 A_o - LA_0 = 0 .$$

In order to treat the boundary conditions we note that $f_{\nu+1}/f_\nu = O(\varepsilon)$ as $\varepsilon \to 0$. Substituting (3.27) into (3.28) and taking only the terms in f_o we obtain

3.32 $$\alpha\, A_o + \beta \frac{\partial A_o}{\partial z} = 0 ,$$

so that $(\nabla\theta)^2 = \lambda_p^2$ as usual with $A_o = a_o \psi_p$.
To obtain the transport equation we need to equate the coefficient of f_{-1} in (3.30) to zero and then take inner products with ψ_p . This leads to

3.33 $$\theta^2[\lambda_p^2\langle\psi_p,A_1\rangle - \langle\psi_p,LA_1\rangle] + \nabla(\theta^2)\cdot\nabla a_o + \nabla(\theta^2)a_o\langle\psi_p,\nabla\psi_p\rangle + 1/2\ \nabla^2(\theta^2)a_o - 2\langle\psi_p,L\psi_p\rangle\ a_o = 0 .$$

The first term of (3.33) yields a boundary term on integration by parts and so does $\langle\psi_p , \nabla\psi_p\rangle$ by (2.38). But these cancel because of the equation obtained by taking terms of order f_{-1} in (2.16). Thus

3.34 $$(\alpha A_1 + \beta \frac{\partial A_1}{\partial z}) f_1 - \beta \nabla Z \cdot \theta \nabla \theta A_o \varepsilon^2 f_{-1} = 0 .$$

But $\varepsilon^2 f_{-1} = -\frac{1}{\theta^2} f_1$ by (3.29a), so that

3.35 $$\alpha A_1 + \beta \frac{\partial A_1}{\partial z} = -\frac{\beta}{\theta} \nabla Z \cdot \nabla \theta A_o \quad \text{on} \quad z = Z^{\pm} .$$

All that remains of (3.33) is the zero order transport equation

3.36 $$\nabla(\theta^2) \cdot \nabla a_o + 1/2 \, \nabla^2(\theta^2) a_o - 2(\nabla\theta)^2 a_o = 0 .$$

This may be put in the form

3.37 $$\frac{\theta^2}{a_o} \nabla \cdot (a_o^2 \frac{\nabla\theta}{\theta}) = 0 .$$

The conservation equation analogous to (2.41) is thus

3.38 $\lambda_p a_o^2 \, \delta\sigma/\theta$ is constant along a ray tube.

But here $\delta\sigma/\theta$ and hence a_o , is finite and non-zero as $\theta \to 0$ at the source. So far so good! The difficulty arises when we now take the inner product of the coefficient of f_{-1} with ψ_m , $m \neq p$, to get $<\psi_m , A_1>$. This leads after some reduction to

3.39 $$\theta^2(\lambda_p^2 - \lambda_m^2) <\psi_m, A_1> + 1/2 \, \nabla(\theta^2)[<\psi_m, \nabla\psi_p> - <\nabla\psi_m, \psi_p>] a_0 = 0 .$$

But as $\theta \to 0$

3.40 $$1/2 \, \nabla(\theta^2) = O(\theta)$$

so that

3.41 $<\psi_m , A_1>$ is $O(\theta^{-1})$

unless

$$3.42 \qquad \langle\psi_m , \nabla\psi_p\rangle - \langle\nabla\psi_m , \psi_p\rangle = 0 .$$

This points up the difficulty and shows that the form (3.27) is not uniform near the source since when θ is sufficiently small the term in f_1 will dominate the one in f_o .

Nevertheless, we shall assume the term in f_o gives a good zero order approximation even though we do not know the correct form for later terms in a uniformly asymptotic series.

In order to find a_o from (3.36) we need some initial conditions. We shall assume that at the source each mode is excited to the same extent that it would be if the medium were perfectly horizontally stratified with properties everywhere the same as at $x = 0$, $y = 0$, the source horizontal coordinates. Thus for a point source at $z = Z_s$ in such a medium we have

$$3.43 \qquad \phi(r,z) = i\pi \sum_{p=0}^{\infty} H_o^{(2)}\left(\frac{\lambda_p r}{\varepsilon}\right) \psi_p (Z_S) \psi_p (z) .$$

Each term in (3.43) which corresponds to a propagating mode will give rise to a ray solution whose leading coefficient will satisfy

$$3.44 \qquad a_0 = -i\pi \psi_p(0,0,Z_S)$$

at the source point. The equation $(\nabla\theta)^2 = \lambda_p^2$ with $\theta = 0$ at $x = 0$, $y = 0$ will determine θ and the rays after which (3.36) with initial conditions (3.44) will give us a_o for each p . If we denote the a_o belonging to λ_p by a_o^p then the solution in a region containing the source will be taken as

$$3.45 \qquad \phi = \sum_{p=0}^{\infty} a_o^p (x,y)\psi_p(x,y,z)H_o^{(2)} (\theta/\varepsilon)$$

in accordance with (3.27) and the discussion above.

4. Space-time rays for more general time dependence

So far we have considered only time harmonic disturbances which are proportional to $e^{i\omega T}$. After cancellation of the exponential factor such signals are governed by the reduced wave equation (2.1) where

4.1 $$k^2(\varepsilon X,\varepsilon Y,Z) = \frac{\omega^2}{c^2(\varepsilon X,\varepsilon Y,Z)} .$$

In this section we shall generalize the ansatz of the earlier sections for more general time dependence and at the same time we may allow the sound speed $c(\varepsilon X,\varepsilon Y,Z,\varepsilon T)$ and the boundaries $Z = Z^{\pm}(\varepsilon X,\varepsilon Y,\varepsilon T)$ to depend weakly upon the time T.

4.1 The ray theory

We start with the full time-dependent wave equation

4.2 $$\frac{1}{c^2}\frac{\partial^2\phi}{\partial T^2} - \frac{\partial^2\phi}{\partial X^2} - \frac{\partial^2\phi}{\partial Y^2} - \frac{\partial^2\phi}{\partial Z^2} = 0 ,$$

with boundary conditions in the simple form

4.3 $$\frac{\partial\phi}{\partial T} = 0 \text{ at } Z = 0 , \frac{\partial\phi}{\partial n} = 0 \text{ at } Z = Z^+(\varepsilon X,\varepsilon Y,\varepsilon T) .$$

We choose this form of boundary condition partly for convenience. It would in principle be possible to allow the upper boundary to be the free surface of the ocean disturbed by surface waves. However, in order to set up the correct equations in that case we should need to linearize the equations about a dynamic state instead of linearizing about equilibrium but this would lead us too far from our theme. Transforming to the contracted variables

4.4 $$x = \varepsilon X,\ y = \varepsilon Y,\ z = Z,\ t = \varepsilon T$$

(4.2) becomes

4.5 $$\frac{1}{c^2}\partial_t^2\phi - \nabla^2\phi + \frac{1}{(i\varepsilon)^2}\partial_z^2\phi = 0 .$$

We now use the new ansatz replacing (2.10), namely

4.6 $$\phi(x,y,z,t) = e^{S(x,y,t)/(i\varepsilon)} \sum_{\nu=0}^{\infty} A_\nu(x,y,z,t)(i\varepsilon)^\nu ,$$

where S is not necessarily linear in t and A_ν may depend upon t as well as x,y,z .

On substituting (4.6) into (4.5) and cancelling the exponential we get

4.7 $$\sum_{\nu=0}^{\infty} \Big\{ \frac{1}{c^2}(S_t^2 A_\nu + S_{tt}A_{\nu-1} + 2S_t A_{\nu-1,t} + A_{\nu-2,tt}) - [(\nabla S)^2 A_\nu + 2\nabla S\cdot\nabla A_{\nu-1} + \nabla^2 A_{\nu-1} + \nabla^2 A_{\nu-2}] + \partial_z^2 A_\nu \Big\} (i\varepsilon)^{\nu-2} = 0 .$$

Equating coefficients of $(i\varepsilon)^{\nu-2}$ to zero starting with $\nu = 0$ we obtain

4.8 $$\frac{1}{c^2} S_t^2 A_o - (\nabla S)^2 A_o + \partial_z^2 A_o = 0;\ A_o = 0,\ z = 0;\ \partial_z A_o = 0,\ z = Z^+ .$$

Let us write

4.9 $$\omega = -S_t \ , \ \underline{k} = \nabla S .$$

Then (4.8) shows that

4.10 $$A_o = a_o \psi_p(\omega,x,y,t) ,$$

where

4.11a $$\partial_z^2 \psi_p + (\frac{\omega^2}{c^2} - \underline{k}^2)\ \psi_p = 0 ,$$

4.11b $$\psi_p = 0,\ z = 0;\ \partial_z \psi_p = 0,\ z = Z^+ ;$$

4.11c $$\langle\psi_p,\psi_q\rangle \equiv \int_0^{Z^+} \psi_p \psi_q dz = \delta_{pq} .$$

For a given ω this is an eigenvalue problem like (2.22), (2.23) but here ω is not given in advance and we must regard (4.10), (4.11) as defining a relation between ω and $\underline{k}$

4.12 $$\underline{k}^2 = \lambda_p^2(\omega,x,y,t) \text{ , say ,}$$

where as in (2.24) the λ_p^2 form a decreasing sequence tending to $-\infty$. Actually we see from (4.9) that (4.12) is a partial differential equation for S analogous to the eikonal (2.26), and like that equation it may be solved by the method of characteristics. The characteristics are defined by the ordinary differential equations ([16])

4.13 $$\frac{dx}{k_x} = \frac{dy}{k_y} = \frac{dt}{\lambda_p \partial_\omega \lambda_p} = \frac{dS}{\lambda_p(\lambda_p - \omega\partial_\omega \lambda_p)} = \frac{dk_x}{\lambda_p \partial_x \lambda_p}$$

$$= \frac{dk_y}{\lambda_p \partial_y \lambda_p} = \frac{-d\omega}{\lambda_p \partial_t \lambda_p} ,$$

and are called <u>horizontal space-time rays</u>. This system may be solved simultaneously for x, y, t, k_x, k_y ω and S provided that starting values are given satisfying (4.12) at some point on the ray. Notice that if c and the boundary conditions are independent of t then so is λ_p and by (4.13) ω is constant along each ray.

On differentiating (4.12) with respect to $\underline{k}$ we obtain

4.14 $$\underline{k} = \lambda_p \partial_\omega \lambda_p \nabla_{\underline{k}} \omega .$$

Thus the first three members of (4.13) give

4.15 $$\frac{d\underline{x}}{dt} = \frac{\underline{k}}{\lambda_p \partial_\omega \lambda_p} = \nabla_{\underline{k}} \omega \quad .$$

But this tells us that the <u>horizontal space-time rays are traced by points traveling with the group velocity</u>. We now equate the coefficient of $(i\varepsilon)^{-1}$ in (4.7) to zero.

This gives

$$4.16 \qquad [\frac{1}{c^2} S_t^2 A_1 - (\nabla S)^2 A_1 + \partial_z^2 A_1] + 2(\frac{1}{c^2} S_t A_{o,t} - \nabla S \cdot \nabla A_o) + (\frac{1}{c^2} S_{tt} - \nabla^2 S) A_o = 0$$

with boundary condition (compare (2.34))

$$4.17 \qquad A_1 = 0 \quad \text{at } z = 0, \ \partial_z A_1 = -\nabla Z^+ \cdot \nabla S \ A_o \ .$$

Taking an inner product of (4.16) with ψ_p we obtain

$$4.18 \qquad <\psi_p, (\frac{\omega^2}{c^2} - \underline{k}^2) A_1 + \partial_z^2 A_1> - 2<\psi_p, \frac{\omega}{c^2}(a_{o,t}\psi_p + a_o \psi_{p,t}) + \underline{k}\cdot\nabla a_o \psi_p + \underline{k}\cdot\nabla\psi_p a_o> - <\psi_p, (\frac{1}{c^2}\omega_t + \nabla\cdot\underline{k}) a_o \psi_p> = 0 \ .$$

In order to reduce (4.18) we use an equation obtained by differentiating (4.11c) with $q = p$:

$$4.19 \qquad <\psi_p, \nabla\psi_p> = -\frac{1}{2}\psi_p^2\Big|_{Z^+} \nabla Z^+$$

and one obtained by differentiating (4.11a) with respect to $\underline{k}$:

$$4.20 \qquad \partial_z^2 \nabla_{\underline{k}} \psi_p + (\frac{\omega^2}{c^2} - k^2)\nabla_{\underline{k}}\psi_p + 2(\frac{\omega}{c^2}\nabla_{\underline{k}}\omega - \underline{k})\psi_p = 0 \ ,$$

which leads on taking an inner product with ψ_p to

$$4.21 \qquad \omega\nabla_k \omega <\psi_p, \frac{1}{c^2}\psi_p> = \underline{k} \ ,$$

or

$$4.22 \qquad \omega|\nabla_k \omega| \ <\psi_p, \frac{1}{c^2}\psi_p> = |\underline{k}| \ .$$

Thus (4.18) becomes

4.23 $$[\psi_p \partial_z A_1 - \partial_z \psi_p A_1]_0^{Z^+} + \psi_p^2\Big|_{Z^+} \underline{k}\cdot\nabla Z^+ a_o - 2\omega<\psi_p,\frac{1}{c^2}\psi_{p,t}>a_o$$

$$- 2[\omega<\psi_p,\frac{1}{c^2}\psi_p>a_{o,t} + \underline{k}\cdot\nabla a_o] - [<\psi_p,\frac{1}{c^2}\psi_p>\omega_t + \nabla\cdot\underline{k}]a_o = 0 .$$

But the first two terms of (4.23) cancel by virtue of (4.11) and (4.17). Thus finally we obtain using (4.21), (4.22)

4.24 $$2\omega<\psi_p,\frac{1}{c^2}\psi_p>(a_{o,t} + \nabla_k\omega\cdot\nabla a_o)$$

$$+ (<\psi_p,\frac{1}{c^2}\psi_p>\omega_t + \nabla\cdot k + 2\omega<\psi_p,\frac{1}{c^2}\psi_{p,t}>)a_o = 0 .$$

This is an ordinary differential equation for a_o along a ray since by (4.15) it may be written

4.25 $$2\omega<\psi_p,\frac{1}{c^2}\psi_p>\frac{da_o}{dt}$$

$$+ (<\psi_p,\frac{1}{c^2}\psi_p>\omega_t + 2\omega<\psi_p,\frac{1}{c^2}\psi_{p,t}> + \nabla\cdot\underline{k})a_o = 0 .$$

We note that if c is independent of t, as is commonly the case, (4.23) on multiplication by a_o gives

4.26 $$\partial_t(\omega<\psi_p,\frac{1}{c^2}\psi_p> a_o^2) + \nabla\cdot(a_o^2\underline{k}) = 0 .$$

This is a space-time divergence equation and by (4.15), (4.21) the space-time vector

4.27 $$(a_o^2\underline{k},\ \omega<\psi_p,\frac{1}{c^2}\psi_p> a_o^2)$$

is parallel to the rays. Thus on integrating (4.26) along a narrow tube of rays bounded by surfaces S_1, S_2 on which t = constant we have

4.28 $$\left[\delta\Sigma\ \omega\langle\psi_p, \frac{1}{c^2}\psi_p\rangle a_o^2\right]_{S_1}^{S_2} = 0 .$$

where $\delta\Sigma$ is the area cut out by the tube of rays on surfaces t = constant. Thus

4.29 $$\delta\Sigma\ \omega\langle\psi_p, \frac{1}{c^2}\psi_p\rangle a_o^2 \text{ is constant along rays.}$$

By (4.22) this reduces to

4.30 $$\frac{\delta\Sigma|\underline{k}|a_o^2}{v_g} = \text{constant along rays,}$$

where

4.31 $$v_g = \frac{\partial\omega}{\partial|k|}$$

is the group velocity.

Just as for the spatial rays of section 2 further consideration is required when $\delta\Sigma$, the area cut out by the tube of rays on surfaces t = constant, goes to zero. This can lead to an 'Airy phase' when the travel-time along rays to a particular location x,y has a (local) maximum or minimum. This is analogous to a smooth caustic. The point source needs special attention and so does the high frequency contribution since when ω is very large all rays travel with almost the same speed c, the characteristic sound speed of the medium on the axis of the sound channel.

4.2 The excitation due to a point source

Just as in section 3.2 we shall assume that near the source the leading term in (4.6) is excited by a point source to the same extent that it would be excited if the medium had no horizontal or temporal variation. The method we use here is different from that of section 3.2 and is an example of the use of canonical problems. This method gives the leading term correctly but would fail for higher terms even if the difficulty mentioned after (3.42) were absent.

We shall solve an inner problem in order to find the behavior of a_o near a point source so as to have starting values for (4.25) or to determine the constant in (4.30).

We consider

4.32 $$\frac{1}{c^2(0,0,Z,0)}\,\partial_T^2\phi - \partial_X^2\phi - \partial_Y^2\phi - \partial_Z^2\phi = f(T)\delta(X)\delta(Y)\delta(Z-Z_S) .$$

$$\phi = 0,\ Z = 0;\ \partial_Z\phi = 0,\ Z = Z^+(0,0,0) .$$

The right member of (4.32) represents the source localized at $(0,0,Z_o)$ and with time variation like $f(T)$. We shall assume that $f(T)$ is zero outside some interval $(0,T_1)$. Notice that c, Z^+ have been specialized by setting $\varepsilon = 0$ in $c(\varepsilon X,\varepsilon Y,Z,\varepsilon T)$, $Z^+(\varepsilon X,\varepsilon Y,\varepsilon T)$.

Let us transform (4.32) by setting

4.33 $$\hat{\phi}_p(\omega,X,Y) = \int_{-\infty}^{\infty} dT \int_{-\infty}^{\infty} dZ\phi(X,Y,Z,T)\psi_p(\omega,Z)e^{-i\omega T} ,$$

where as usual

4.34 $$\partial_Z^2\psi_p + (\frac{\omega^2}{c^2} - \lambda_p^2)\psi_p = 0 ,$$

$$\psi_p = 0,\ Z = 0;\ \partial_Z\psi_p = 0,\ Z = Z^+(0,0,0) .$$

Transforming equation (4.32) we obtain

4.35 $$-\int_{-\infty}^{\infty} \langle\psi_p,(\frac{\omega^2}{c^2} - \partial_Z^2)\phi_p\rangle e^{-i\omega T} dT - \partial_X^2\hat{\phi} - \partial_Y^2\hat{\phi} = \hat{f}(\omega)\delta(X)\delta(Y)\psi_p(Z_S)$$

which leads by way of (4.34) to

4.36 $$\partial_X^2\hat{\phi}_p + \partial_Y^2\hat{\phi}_p + \lambda_p^2\hat{\phi}_p = -\hat{f}(\omega)\delta(X)\delta(Y)\psi_p(Z_S) .$$

Here $\hat{f}(\omega)$ is the Fourier transform of $f(T)$.

As in section 3.2

4.37 $$\hat{\phi}_p = - i\pi\hat{f}(\omega)\psi_p(Z_S)H_o^{(2)}(\lambda_p R)$$

which satisfies the outgoing radiation condition at $R = \infty$, where

4.38 $$R = (X^2 + Y^2)^{\frac{1}{2}} .$$

Thus a good approximation to ϕ is

4.39 $$\phi(X,Y,Z,T) = \sum_p \phi_p(X,Y,Z,T) ,$$

where

4.40 $$\phi_p(X,Y,Z,T) = - \frac{i}{2}\int_{-\infty}^{\infty} \hat{f}(\omega)\psi_p(\omega,Z_S)H_o^{(2)}[\lambda_p(\omega)R]\psi_p(\omega,Z)e^{-i\omega T}d\omega .$$

The outer expansion of this inner solution is obtained by setting

4.41 $$R = r/\varepsilon , \quad r = (x^2 + y^2)^{\frac{1}{2}} , \quad T = t/\varepsilon$$

and evaluating asymptotically for small ε . Thus

4.42 $$\phi_p(x,y,z,t) = - \frac{i}{2}\int_{-\infty}^{\infty} \hat{f}(\omega)\psi_p(\omega,Z_S)\psi_p(\omega,z)H_o^{(2)}\left(\frac{\lambda_p r}{\varepsilon}\right)e^{i\omega t/\varepsilon}\, d\omega$$

Using the leading term of the asymptotic series for $H_o^{(2)}$ with large argument ([17]) we obtain

4.43 $$\phi_p(x,y,z,t) \sim$$

$$- \frac{i}{2}\int_{-\infty}^{\infty} \hat{f}(\omega)\psi_p(\omega,Z_S)\psi_p(\omega,z)\left(\frac{2\varepsilon}{\pi\lambda_p r}\right)^{\frac{1}{2}} e^{i\frac{\pi}{4}} e^{[\lambda_p(\omega)r-\omega t]/(i\varepsilon)} d\omega .$$

But this integral is suitable for an application of the stationary phase approximation. Let $\omega_o = \omega_o(r,t)$ be a value of ω for which

4.44
$$\partial_\omega[\lambda_p(\omega)r - \omega t]\Big|_{\omega_o} = 0 .$$

Then, making the usual approximations near ω_o when only the leading term is required, we set

4.45
$$\lambda_p(\omega)r-\omega t \sim \lambda_p(\omega_o)r-\omega_o t + \frac{1}{2}\,\partial^2_\omega\lambda_p(\omega_o)r(\omega-\omega_o)^2$$

in the exponential but $\omega = \omega_o$ in the other factors. This approximation yields

4.46
$$\begin{aligned}\phi_p(x,y,z,t) &\sim -\frac{i}{2}\,\hat{f}(\omega_o)\psi_p(\omega_o,Z_S)\psi_p(\omega_o,z)\\ &\quad\times e^{i\frac{\pi}{4}}\left(\frac{2\varepsilon}{\pi\lambda_p(\omega_o)r}\right)^{\frac{1}{2}} e^{[\lambda_p(\omega_o)r-\omega_o t]/(i\varepsilon)}\\ &\quad\times \int_{-\infty}^{\infty} e^{\frac{1}{2}\partial^2_\omega\lambda_p(\omega_o)(\omega-\omega_o)^2/(i\varepsilon)}\,d\omega\\ &\sim e^{\frac{\pi}{4}i(1\pm 1)}\,\frac{1}{[\lambda_p(\omega_o)|\partial^2_\omega\lambda_p(\omega_o)|]^{1/2}}\,\frac{\varepsilon}{r}\\ &\quad\times \hat{f}(\omega_o)\psi_p(\omega_o,Z_S)\psi_p(\omega_o,z)e^{[\lambda_p(\omega_o)r-\omega_o t]/(i\varepsilon)} ,\end{aligned}$$

where the upper or lower sign is to be used according as $\partial^2_\omega\lambda_p(\omega_o)$ is positive or negative.

Equation (4.46) is now suitable for comparison with (4.6) when x,y,t are small. However, we must make one minor modification to (4.6): it should first be multiplied by ε . This change makes no difference to the analysis in the previous section and we see that, as $r \to 0$, $t \to 0$ in such a way that r/t is constant, we get

4.47 $$\lim_{r\to 0} rA_o = \frac{e^{\frac{3}{4}i(1\pm 1)}}{[\lambda_p(\omega_o)|\partial^2_\omega\lambda_p(\omega_o)|]^{1/2}}\hat{f}(\omega)\psi_p(\omega_o,Z_S)\psi_p(\omega_o,z) ,$$

4.48 $$S(x,y,t) \sim \lambda_p(\omega_o)r-\omega_o t ,$$

where ω_o is given in terms of r,t by (4.44). It is easily seen that in (4.30) $\delta\Sigma$ is $O(r^2)$ as $r \to 0$ when all rays issue from one point. Hence by (4.47) the constant in (4.30) is well determined as $r \to 0$.

The procedure by which the leading term is determined is as follows. Given the point x,y,t and mode number p, find the space-time horizontal rays for mode p which join (0,0,0) to (x,y,t) . For each such ray the value ω_o of ω and the ratio r/t is well determined as $r \to 0$ and moreover these values are consistent with (4.44) since the rays are traced by points moving with the group velocity at each point. Starting values of a_o may now be obtained from (4.47) for use in (4.25) or (4.30) from which a_o may be calculated at each point on the ray and in particular at (x,y,t) .

We note that when ω is large $\lambda_p(\omega)$ frequently has the asymptotic behavior

4.49 $$\lambda_p(\omega) = n_o\omega + n_1 + n_2\omega^{-1} + O(\omega^{-1}) .$$

Thus as $t/r \to n_o$, ω_o of (4.44) tends to infinity and $|\partial^2_\omega\lambda_p(\omega_o)| \to 0$ so that the approximation (4.46) is useless. We shall consider this situation in section 4.4.

4.3 The Airy phase

It frequently happens that in stratified media the group velocity has a minimum for some value of the frequency ([1]). Let ω_o be this frequency and $t_o(x,y)$ the arrival time at (x,y) corresponding to a disturbance from a point source at the origin traveling with this minimum group velocity. Then for $t < t_o$ and $|t-t_o|$ small, there are two space-time rays which arrive at (x,y,t) with slightly differing values of ω . When $t = t_o$ just one ray arrives and when $t > t_o$ no rays reach (x,y,t) . This phenomenon may be generalized for almost stratified media and is the space-time analog of a smooth caustic. For any given location x, y there may be a $t_o(x,y)$ such that two rays join (0,0,0) to (x,y,t) if $t < t_o(x,y)$ and none if $t > t_o(x,y)$. For t near $t_o(x,y)$ we need a more sophisticated ansatz than (4.6). The correct form is a generalization of (3.1).

4.50 $$\phi(x,y,z,t) = e^{S(x,y,t)/i\varepsilon} \sum_{\nu=0}^{\infty} (i\varepsilon)^{\nu}\{A_{\nu}(x,y,t)V[\varepsilon^{-\frac{2}{3}}\rho(x,y,t)] + i\varepsilon^{\frac{1}{3}}B_{\nu}(x,y,t)V'[\varepsilon^{-\frac{2}{3}}\rho(x,y,t)]\} .$$

The calculation proceeds as before. Substituting into (4.5) and equating coefficients of $(i\varepsilon)^{\nu}V$, $(i\varepsilon)^{\nu}i\varepsilon^{1/3}V'$ separately to zero, we get first the eigenvalue problem

4.51 $$\partial_z^2 A_o + [\frac{1}{c^2}(S_t^2+\rho\rho_t^2) - (\nabla S)^2 - \rho(\nabla\rho)^2]A_o = 0$$

where instead of (3.8) we have imposed

4.52 $$\frac{1}{c^2}\rho_t S_t - \nabla\rho\cdot\nabla S = 0 .$$

A_o satisfies the usual boundary conditions

4.53 $$A_o = 0 \text{ at } z = 0, \frac{\partial A_o}{\partial z} = 0 \text{ at } z = Z^+(x,y,t) ,$$

similar equations hold for B_o . Thus as before we write

4.54 $$A_o = a_o^p\psi_p , B_o = b_o^p\psi_p .$$

The transport equations obtained from the $\nu = 1$ coefficients are

4.55a $$2[<\psi_p,\frac{1}{c^2}\psi_p>S_t a_{o,t} - \nabla S\cdot\nabla a_o] + [<\psi_p,\frac{1}{c^2}\psi_p>S_{tt} - \nabla^2 S] a_o + \rho[<\psi_p,\frac{1}{c^2}\psi_p>\rho_{tt} - (\nabla\rho)^2]b_o + [<\psi_p,\frac{1}{c^2}\psi_p>\rho_t^2 - (\nabla\rho)^2] b_o$$

$$+ 2\rho[<\psi_p, \frac{1}{c^2}\psi_p>\rho_t b_{o,t} - (\nabla\rho)^2]\, b_o = 0 ,$$

4.55b $$2[<\psi_p,\frac{1}{c^2}\psi_p>S_t b_{o,t} - \nabla S\cdot\nabla b_o] + [<\psi_p,\frac{1}{c^2}\psi_p>S_{tt} - \nabla^2 S]\, b_o$$

$$+ [<\psi_p,\frac{1}{c^2}\psi_p>\rho_{tt} - (\nabla\rho)^2]a_o + 2[<\psi_p,\frac{1}{c^2}\psi_p>\rho_t a_{o,t} - \nabla\rho\cdot\nabla a_o] = 0 .$$

Just as at (3.13-3.17) it is found that on defining

4.56 $$S^{\pm} = S \pm \frac{2}{3}\rho^{\frac{3}{2}} ,$$

$$a_o^{\pm} = \rho^{-\frac{1}{4}} a_o \pm \rho^{\frac{1}{4}} b_o ,$$

that $a_o^{\pm}$ satisfy the ordinary transport equation (4.24) and $S^{\pm}$ the ordinary eikonal equation (4.12), (4.9).

As we mentioned earlier this ansatz is suitable for the case where two rays reach some points (x,y,t) . For such points $S^{\pm}$ are the values of S which would be computed from (4.13) for the two rays, and $\rho(x,y,t) = 0$ defines the surface $t = t_o(x,y)$ separating the region reached by two rays from the region reached by no rays.

The following method suggests itself for implementing the ansatz (4.50). Let us assume that we have at our disposal a computer ray tracing code and a method of solving the ordinary transport equations. Then let S^- be the phase at (x,y,t) corresponding to the ray which has not yet touched the caustic, let S^+ be the phase at (x,y,t) on the ray which has touched the caustic. Then define

4.57 $$S = \frac{1}{2}(S^+ + S^-) ,$$

$$\rho = [\tfrac{3}{4}(S^+ - S^-)]^{\frac{2}{3}} .$$

Let a_o^- be the amplitude on the ray which has not touched the caustic calculated in the usual way and let a_o^+ be the amplitude on the other ray continued beyond the caustic by means of (4.30) where $\delta\Sigma$ is always taken to be positive. Then define

4.58 $$a_o = \frac{1}{2}\rho^{\frac{1}{4}}(a_o^+ + a_o^-)$$

$$b_o = \frac{1}{2}\rho^{-\frac{1}{4}}(a_o^+ - a_o^-) .$$

If the calculation is accurate it will be found that a_o , b_o are finite as $\rho \to 0$ although $a_o^{\pm}$ both become infinite, and ρ itself is a smooth function tending to zero as the field point approaches the caustic. With these values of S, ρ , a_o , b_o we may use the first term $\nu = 0$ in (4.50) to obtain an approximation valid right up to the caustic.

This calculation depends upon the existence of two rays reaching the point (x,y,t) and so it fails on the dark side of the caustic where no rays penetrate. However, the required values of S, ρ in the neighborhood of the caustic but on the dark side may be obtained by extrapolation from the bright side. Indeed, we never need extrapolate far since the exponentially decreasing behavior of $V(\zeta)$ for negative argument assures us that only small negative values of ρ are significant.

It is of interest to see how the ansatz (4.49) reduces to the original form (4.6) when ρ is not small. In this case we may approximate $V(\varepsilon^{-2/3}\rho), V'(\varepsilon^{-2/3}\rho)$ using the large negative argument approximations to the Airy function ([18]) :

4.59 $$V(\zeta) = Ai(-\zeta) = \frac{1}{\sqrt{\pi}}\zeta^{-\frac{1}{4}}\sin(\frac{2}{3}\zeta^{\frac{3}{2}} + \frac{1}{4}\pi) + O(\zeta^{-\frac{1}{4}}) ,$$

$$V'(\zeta) = - Ai'(-\zeta) = \frac{1}{\sqrt{\pi}}\zeta^{\frac{1}{4}}\cos(\frac{2}{3}\zeta^{\frac{3}{2}} + \frac{1}{4}\pi) + O(\zeta^{-\frac{5}{4}}) .$$

Thus by (4.49)

$$4.60 \qquad \phi \sim e^{S/i\varepsilon}\{\frac{1}{\sqrt{\pi}}\varepsilon^{\frac{1}{6}}\rho^{-\frac{1}{4}}a_o\sin(\frac{2}{3}\frac{\rho^{3/2}}{\varepsilon}+\frac{1}{4}\pi)$$

$$+\frac{i}{\sqrt{\pi}}\varepsilon^{\frac{1}{6}}\rho^{\frac{1}{4}}b_o\cos(\frac{2}{3}\frac{\rho^{3/2}}{\varepsilon}+\frac{1}{4}\pi)\}\rho_p$$

$$\sim \frac{1}{2\sqrt{\pi}}\varepsilon^{\frac{1}{6}}e^{S/i\varepsilon}\{a_o^- e^{i\rho^{3/2}/\varepsilon}e^{-\frac{1}{4}\pi i}+a_o^+ e^{-i\rho^{3/2}/\varepsilon}e^{\frac{1}{4}\pi i}\}\psi_p$$

$$\sim \frac{1}{2\sqrt{\pi}}\varepsilon^{\frac{1}{6}}\{a_o^- e^{S^-/i\varepsilon}e^{-\frac{1}{4}\pi i}+a_o^+ e^{S^+/i\varepsilon}e^{\frac{1}{4}\pi i}\}\psi_p .$$

This agrees with (4.6) if we divide (4.59) by $\varepsilon^{\frac{1}{6}}$, multiply by $2\sqrt{\pi}\, e^{(1/4)\pi i}$ and identify a_o^- of (4.60) with a_o of (4.6), (4.10).

We see from (4.60) that there is a phase shift of $\frac{\pi}{2}$ in the signal corresponding to the ray which has touched the caustic relative to the signal for the direct ray even after allowance has been made for the differences in path length. Provided this phase shift is incorporated it is possible to use the naive ray theory of (4.6) even for rays which have grazed the caustic, provided the field point is not close to the caustic. The ansatz (4.50), from which (4.59) was derived, provides a connection formula for rays which graze the caustic.

Equation (4.39) also shows that the field away from the caustic is smaller by order of magnitude $\varepsilon^{1/6}$ than the field near the caustic. Or, looking at the phenomenon the other way, the field at the caustic is amplified by a factor of order $\varepsilon^{-(1/6)}$ relative to the general field away from the caustic. Indeed, it is this intensification which gives caustics their name. In the time dependent case the Airy phase manifests itself as a large amplitude oscillation which terminates a dispersed train of oscillations consisting of two superposed frequencies. The low frequency component increases in frequency and the high frequency component decreases in frequency until they terminate with a common frequency in the Airy phase which with its large amplitude is often the most prominent feature on the record ([1]).

4.4 The precursor and other phenomena requiring special treatment

As we mentioned at the end of section 4.4 special consideration needs to be given to the high frequency arrivals which propagate with speeds close to the characteristic speed on the axis of the wave guide. For some velocity profiles (fairly flat ones) this speed is approached from below as $\omega \to \infty$ while for others (with deep velocity minima on the axis) the velocity may approach c from above as $\omega \to \infty$. In either case the dispersion relation typically has the asymptotic form

$$4.61 \qquad \lambda_p(\omega) = n_o\omega + n_1 + n_2\omega^{-1} + O(\omega^{-1})$$

It will be seen that when t/r is near n_o in (4.44) the frequency ω_o will be large and $\partial^2_\omega \lambda_p$ in (4.46) will be small so that the approximation used there is not valid. Moreover all modes have the same group velocity in this high frequency limit so that rays corresponding to different modes tend to arrive simultaneously. But under these conditions the parabolic equation method as developed by Tappert [19] is probably the most useful method. This high frequency arrival is usually referred to as the water wave ([1]).

If, with the advent of arrays of receivers which can separate individual modes, a theory for the water wave carried by an individual mode is required, it may be possible to adapt the method of Zauderer [20]. If the time function f(T) in (4.32) is replaced by $\delta(T)$ so that we are seeking a fundamental solution then the equation in the x, y, z, t variables is

$$4.62 \qquad \frac{1}{c^2}\,\partial_t^2\phi - \nabla^2\phi - \frac{1}{\varepsilon^2}\,\partial_z^2\phi = \varepsilon\,\delta(t)\,\delta(x)\,\delta(y)\,\delta(z-z_o)\,.$$

Then Zauderer's prescription would suggest we set

$$4.63 \qquad \phi(x,y,z,t) = \sum_{\nu=0}^{\infty} A_\nu(x,y,z,t)\; g_\nu[\varepsilon,S(x,y,t)]$$

where

$$4.64 \qquad g_\nu(\varepsilon,S) = \varepsilon(\varepsilon S)^{\nu-1/2}\, J_{\nu-1/2}(S/\varepsilon)\ ,$$

which should be compared with (3.28). Then

4.65 $$\frac{1}{\varepsilon^2} g_\nu = (2\nu-3)g_{\nu-1} - \theta^2 g_{\nu-2}$$

$$\partial_\theta g_\nu = \theta g_{\nu-1}$$

which are similar to (3.29a) and (3.29b). On substituting (4.63) into (4.62) we obtain the usual eikonal eigenvalue problem

4.66 $$\partial_z^2 A_o + (\frac{\omega^2}{c^2} - \underline{k}^2)A_o = 0 ,$$

where $\omega = - S_t$, $\underline{k} = \nabla S$ so that $A_o = a_o \psi_p$. The transport equation for a_o is

$$[<\psi_p, \frac{1}{c^2} \psi_p>(S^2)_t a_{o,t} - \nabla(S^2)\cdot\nabla a_o] + <\psi_p, \frac{1}{c^2} \psi_{pt}>(S^2)_t a_o$$

$$+ \frac{1}{2}[<\psi_p, \frac{1}{c^2} \psi_p>(S^2)_{tt} - \nabla^2(S^2)]a_o + 3<\psi_p, \partial_t^2\psi_p>a_o = 0 .$$

If c is independent of t this may be put into the form

4.67 $$\partial_t(<\psi_p, \frac{1}{c^2} \psi_p>S^{-2}S_t a_o{}^2) - \nabla\cdot(S^{-2}\nabla S a_o^2) = 0 ,$$

which gives a finite value of a_o at the source. Compare (3.37). Just as (4.26) leads to (4.30) so (4.64) leads to

4.68 $$\frac{\delta\Sigma\ |\underline{k}|}{S^2 v_g} a_o^2 = \text{constant along a ray,}$$

where $\underline{k}$, $\delta\Sigma$, v_g have the same significance as in (4.30). Starting values for a_o near the origin may be obtained by solving an inner problem using the method of Handelsman and Bleistein [21] to evaluate (4.43) as t/r approaches n_o of (4.61). In order to do this the constants in (4.61) must be known. We shall not pursue this topic further.

Another phenomenon requiring special treatment is cut-off. It frequently occurs that in a perfectly stratified medium waves fail to progagate for $|\underline{k}|$ and ω below some finite nonzero values ([1]). This is connected with what are often called ground waves in ocean acoustics or head, or lateral, waves in other contexts. They correspond to disturbances traveling in the substratum below the ocean and subsequently being refracted back into the ocean at the critical angle. Values of $|k|$ and ω smaller than their values at cut-off will correspond to modes which leak energy into the substratum and so are evanescent in the horizontal direction. The horizontal ray theory has not been developed for modes traveling near cut-off and since all modes typically have the same group velocity at cut-off it is doubtful if a normal mode theory is adequate. Reference should be made to Cerveny and Ravindra [22] for an account of head waves in seismology.

This ends our discussion of the theoretical aspects of horizontal rays. In the next two sections we illustrate the theory for time harmonic disturbances. In Section 5 two idealized examples are considered and then in Section 6 we treat wave propagation in a realistic ocean and compare the predictions of our theory with observational data.

5. Two theoretical examples

5.1 Homogeneous medium, one free horizontal boundary, one rigid boundary with small constant slope

As a first illustration of the asymptotic technique of Sections 2 and 3 let us consider a model ocean in which the sound speed is constant, the surface Z = 0 is free and the bottom is rigid with a small constant slope. Thus k^2 of (2.1) is a constant. The boundaries are

5.1 $$Z = Z^-(\varepsilon X, \varepsilon Y) = 0$$

$$Z = Z^+(\varepsilon X, \varepsilon Y) = \varepsilon Y ,$$

which lead to

5.2 $$z = Z^{-}(x,y) = 0 \ ,$$
$$z = Z^{+}(x,y) = y \ .$$

The boundary conditions are

5.3 $$\phi = 0 \quad \text{on} \quad z = 0 \ ,$$
$$\frac{\partial \phi}{\partial n} = 0 \quad \text{on} \quad z = y \ .$$

The operator L is given by

5.4 $$L\phi \equiv \frac{\partial^2 \phi}{\partial z^2} + k^2 \phi \ , \quad k^2 \quad \text{constant.}$$

The eigenvalues and eigenfunctions of L are

5.5 $$\lambda_p^2 = k^2 - \frac{(p-1/2)^2 \pi^2}{y^2}$$

and

5.6 $$\psi_p(x,y;z) = \sqrt{\left(\frac{2}{y}\right)} \sin \left[\frac{(p-1/2)\pi z}{y}\right] \ .$$

Since λ_p is independent of x, the first ray equation (2.27a) may be integrated to give

5.7 $$\lambda_p \cos \xi = \lambda_p^o \ ,$$

where $\cos \xi = dx/ds$ so that ξ is the angle between the ray and the x axis. The constant λ_p^o is given by

5.8 $$\lambda_p^o = \lambda_p(0,y_S) \cos \xi_S \ ,$$

where ξ_S is the angle ξ at the source $x = 0$, $y = y_S$. On using the relation $\tan \xi = dy/dx$ we obtain also

$$5.9 \qquad x = \int_{y_S}^{y} \frac{\lambda_p^o}{(\lambda_p^2-\lambda_p^{o2})^{1/2}}\, dy'$$

$$= \pm \frac{\lambda_p^o}{(k^2-\lambda_p^{o2})} \left[\left\{ y'^2(k^2-\lambda_p^{o2}) - (p-1/2)^2\pi^2 \right\}^{1/2} \right]_{y'=y_S}^{y'=y} .$$

The phase θ with initial condition $\theta = 0$ is given by

$$5.10 \qquad \theta = \int_{y_S}^{y} \frac{\lambda_p^2}{(\lambda_p^2-\lambda_p^{o2})^{1/2}}\, dy'$$

$$= (p-1/2)\pi \left[\frac{k^2}{k^2-\lambda_p^{o2}} \; \frac{(\lambda_p^2-\lambda_p^{o2})^{1/2}}{(p-1/2)\pi} y' - \tan^{-1}\left\{ \frac{(\lambda_p^2-\lambda_p^{o2})^{1/2}}{(p-1/2)\pi} y' \right\} \right]_{y'=y_S}^{y'=y}$$

A quantity which we call the ray-bundle aperture will be used in the numerical scheme which finds the amplitudes a_o . The ray-bundle aperture is σ where $\sigma\, \delta\xi_S/\lambda_p$ is the cross-section of the tube of rays which leaves the source in directions between ξ_S and $\xi_S + \delta\xi_S$. It is given by

$$5.11 \qquad |\sigma| = \lambda_p \left. \frac{\partial x}{\partial \xi_S} \right|_y \sin \xi$$

$$= \left. \sqrt{\lambda_p^2 - \lambda_p^{o2}} \right|_S \left. \sqrt{\lambda_p^2 - \lambda_p^{o2}} \right|_R \left[\left| \frac{(\lambda_p^2(k^2-\lambda_p^{o2})-2k^2\lambda_p^{o2})y}{(k^2-\lambda_p^{o2})\sqrt{\lambda_p^2 - \lambda_p^{o2}}} \right| \right]_S^R ,$$

where the subscripts S and R stand for the source $(0,y_S)$ and receiver (x,y) respectively.

Figures 1 through 3 show results for a particular numerical example:

$$5.12 \qquad k = 2\pi ,$$

$$Y_S = 500 ,$$

$$\varepsilon = .01 .$$

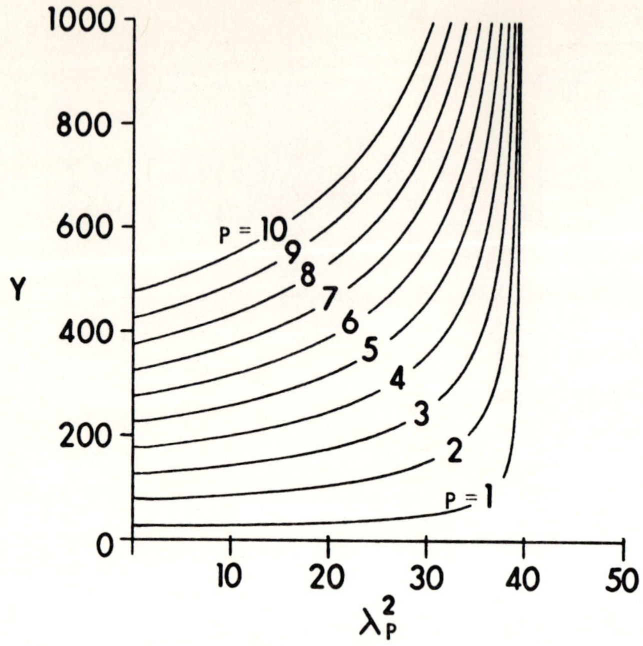

Figure 1. Eigenvalues λ_p^2 versus distance Y from the shore for a homogeneous medium bounded above by a horizontal free surface and below by a rigid surface of constant slope based on (5.5).

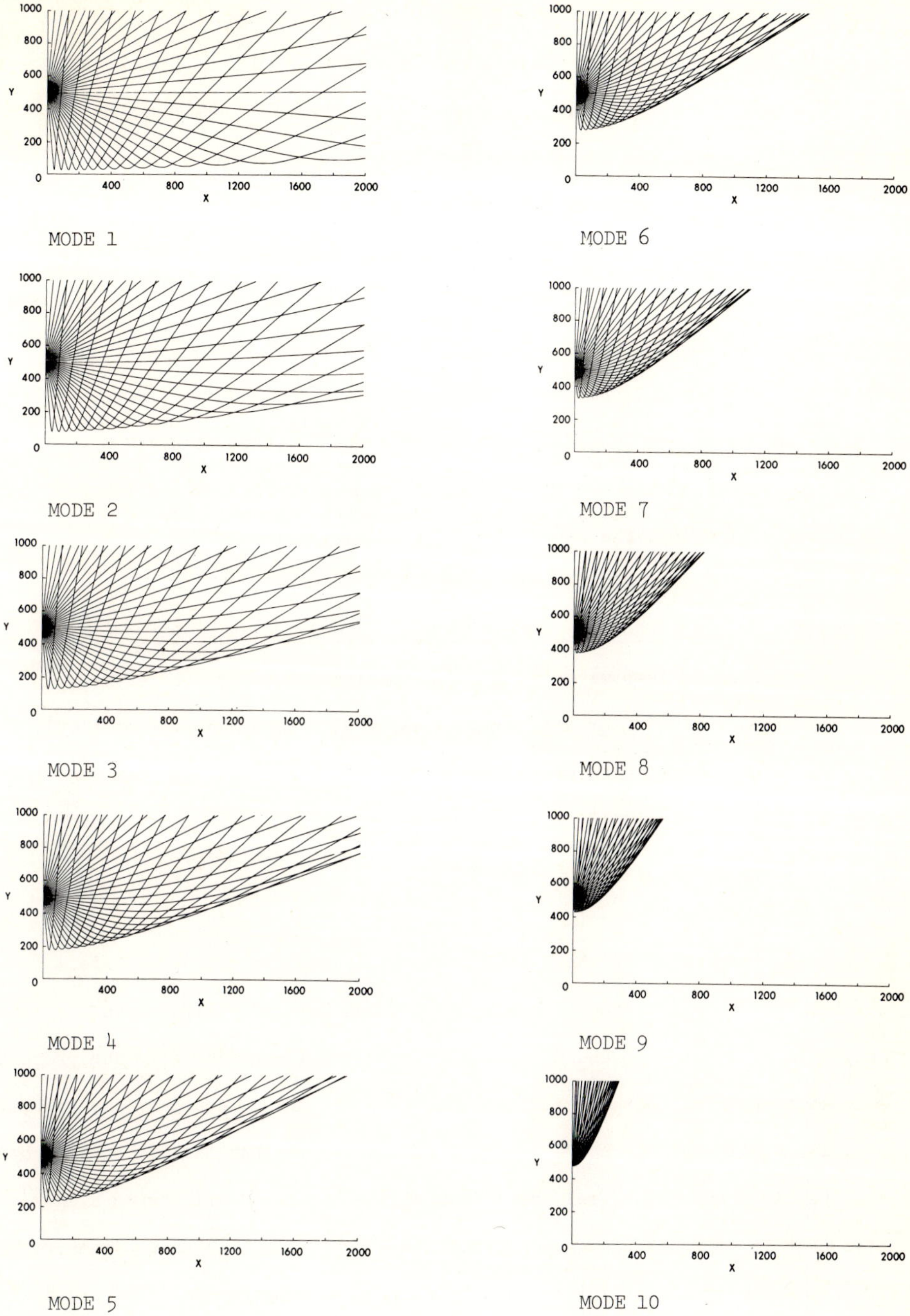

Figure 2. Horizontal ray diagrams for modes 1 through 10 based on (5.9).

It follows from (5.5) that there are only ten propagating modes in the vicinity of the source. Their eigenvalues λ_p^2 (p=1,...,10) are plotted as functions of Y in figure 1. The resulting ray diagrams are plotted in figure 2. Notice that each mode is turned away from the shore Y = 0 (where $Z^+ = Z^- = 0$), and each mode envelops a caustic curve. The caustics corresponding to these ten modes are plotted in figure 3 and labeled with the corresponding mode numbers.

Points on the concave side of curve p but on the convex side of curve p + 1 will receive two rays for each of modes 1,2,...,p and no rays for modes p+1,...,10. Points very near curve p will receive a large amplitude for mode p. However, according to (3.43) the amplitude of mode p excited by a point source at depth Z_S will be proportional to $\psi_p(0,Y_S;Z_S)$. Thus the depth of the source as well as the position of the receiver will affect the amplitude of each mode.

5.2 Propagation in deep water for which the sound speed increases with depth

There are bodies of deep water such as the Mediterranean Sea in February in which the velocity of sound increases monotonically with depth and varies slowly with horizontal position. As a model for such a medium we take k^2 in the form

$$k^2(x,y,z) = k_0^2(x,y) - k_1^2(x,y)z \,. \tag{5.13}$$

In any real body of water the z coordinate will not have the full range $(0,k_0^2/k_1^2)$ but will be restricted to lie between 0 and $Z^+(x,y)$, say, where $z = Z^+(x,y)$ is the equation of the bottom and $Z^+ \ll k_0^2/k_1^2$. We shall consider only the propagation of modes trapped so near the surface that they are not affected by the bottom. The bottom is usually an absorptive boundary so that we may rationalize further by supposing that if a mode does feel the effect of the bottom it will be so highly attenuated that it will not propagate to any great horizontal distance. Thus we shall seek eigenfunctions $\psi_p(x,y,z)$ which ultimately decay as z increases. The eigenvalue problem is now

$$\frac{\partial^2\psi}{\partial z^2} + (k^2-\lambda^2)\,\psi = 0 \,, \tag{5.14}$$

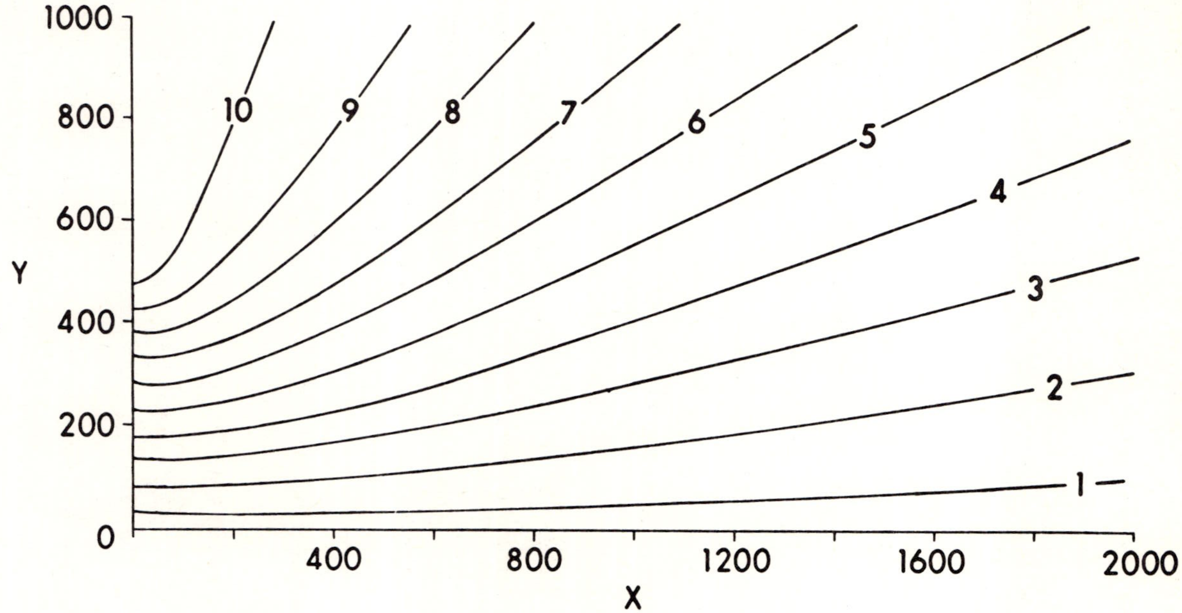

Figure 3. Caustics of propagating modes p = 1,2, ..., 10 which arise because the rays are refracted away from the shoreline.

with

5.15 $\psi = 0$ at $z = 0$

and

5.16 ψ remains bounded.

If in (5.14) we set

5.17 $$\zeta = -(k^2-\lambda^2)/k_1^{4/3}$$

we obtain

5.18 $$\frac{\partial^2\psi}{\partial\zeta^2} = \zeta\psi .$$

Thus if we invoke (5.16) we find that

5.19 $$\psi = c\ \mathrm{Ai}(\zeta) ,$$

where Ai is the Airy function.

Equation (5.15) implies that

5.20 $$\mathrm{Ai}[-(k_o^2-\lambda^2)/k_1^{4/3}] = 0 .$$

Thus λ may take on any of the values λ_p for which

5.21 $$-(k_o^2-\lambda_p^2)/k_1^{4/3} = \zeta_p \text{ , the p-th zero of Ai .}$$

We note that $0 > \zeta_1 > \zeta_2 > \ldots$.

The normalizing constant c of (5.19) may be easily verified to be

5.22 $$c_p = k_1^{1/3}\left[\int_{\zeta_p}^{\infty} \mathrm{Ai}^2(\zeta)d\zeta\right]^{-1/2} .$$

On combining (5.19) with (5.22) we obtain

5.23 $$\psi_p = k_1^{1/3} \left[\int_{\zeta_p}^{\infty} Ai^2 (\zeta) d\zeta \right]^{-1/2} Ai(\zeta_p + k_1^{2/3} Z) .$$

It is interesting that the constant k_o^2 has dropped out of the expression for ψ_p .

Rewriting (5.21) we get

5.24 $$\lambda_p^2 = k_o^2 + k_1^{4/3} \zeta_p .$$

Thus if k_o^2 and $k_1^{4/3}$ depend linearly upon y but not at all on x then so does λ_p^2. In this case the eikonal equation and the ray equation can be integrated exactly.

For example if

5.25 $$\lambda_p^2 = \mu_p^2 + \nu_p^2 y$$

then

5.26 $$x = x_S \pm 2 \frac{\mu_p}{\nu_p} (y^{1/2} - y_S^{1/2})$$

where x_S, y_S are the coordinates of the source,

5.27 $$\theta = \pm \frac{2}{3\nu_p} \left[y^{1/2} (3\mu_p^2 + \nu_p^2 y) - y_S^{1/2} (3\mu_p^2 + \nu_p^2 y_S) \right]$$

and the ray bundle aperture is

5.28 $$\sigma = \mp y_S^{1/2} y^{1/2} \left[\frac{-\mu_p^2 + \nu_p^2 y}{y^{1/2}} - \frac{-\mu_p^2 + \nu_p^2 y_S}{y_S^{1/2}} \right] .$$

The ray diagrams have the same character as those of the preceding example (figure 2.)

6. Long range acoustic propagation in a deep ocean

In this section we show how the asymptotic theory of sections 2 and 3 may be applied to a realistic ocean model. This is made possible by the existence of measurements made during the last decade which give good quantitative information about the variation of sound speed in the ocean, and by the availability of computers which make the computations feasible. We begin by considering from a more practical point of view the parameters which affect acoustic propagation in the (real) ocean. Then we give a brief description of the computer program which implements our scheme. Finally some results for a particular set of sound speed data are presented and these are shown to compare well with observed acoustic amplitudes.

6.1 Environmental parameters

The most important parameter affecting sound propagation in the ocean is the sound speed as a function of position. Empirical formulas such as Leroy's [23] (quoted below) indicate that sound speed increases with temperature, salinity, and depth. Leroy's formula is as follows:

6.1 $$c = c_o + c_a + c_b + c_c + c_d$$

where

(6.2a) $$c_o = 1493 + 3(T-10) - 6\times 10^{-3}(T-10)^2 - 4\times 10^{-2}(T-18)^2 + 1.2(S-35) - 10^{-2}(T-18)(S-35) + 10^{-3}\zeta/61$$

(6.2b) $$c_a = 10^{-1}\zeta^2 + 2\times 10^{-4}\zeta^2(T-18)^2 + 10^{-1}\zeta\,\phi/90 \,,$$

(6.2c) $$c_b = 2.6\times 10^{-4}\,T\,(T-5)\,(T-25)\,,$$

(6.2d) $$c_c = -\,10^{-3}\,\zeta^2\,(\zeta-4)\,(\zeta-8)\,,$$

(6.2e) $$c_d = 1.5\times 10^{-3}\,(S-35)^2(1-\zeta) + 3\times 10^{-6}\,T^2(T-30)\,(S-35)\,,$$

and

c is the sound speed in m/s

ζ is the depth in Km ,

S is the salinity in parts per thousand (by weight) ,

T is the temperature in degrees centigrade ,

ϕ is the latitude in degrees.

The last term in (6.2a) is a corrective term for low salinities, and should not be used if S is greater than 30. Since the dependence upon salinity is slight and the water at great depths is almost isothermal, there is a point below which the sound speed increases almost linearly with depth (See figure 4). The layer immediately above this deep isothermal region is called the main thermocline but occasionally the main thermocline is absent, as in one of the idealized models of section 5.

If the water near the surface is well mixed a surface duct may be formed (see figure 4) but acoustic energy traveling in such a duct tends to be scattered by surface roughness and may not be significant at long ranges.

Figure 5 shows the sound speed along a track in the Pacific Ocean running northward from Hawaii to Alaska during the late summer of 1968. The change in depth of the SOFAR axis, that is, the depth of minimum sound speed at 42° N, is where the Kuroshio and the Oyashio currents meet. Even there the horizontal gradient of sound speed is so small that the asymptotic technique will be applicable.

Up to now we have neglected absorption of energy but to make predictions we must consider the possibility of absorption both in the water itself and at the ocean boundaries. There is currently some disagreement about which empirical absorption formulas are the most accurate. Moreover, much of the relevant data is classified. However, measurements taken by Adlington [24] indicate that the ocean surface acts as a perfect reflector at frequencies below 1 kHz and for wind speeds below 20 knots. Absorption at the ocean bottom, on the other hand, may be quite large. Figure 6 shows the bottom loss (ratio of incident to reflected intensity in decibels) measured by Marsh [25,26] as a function of grazing angle. This refers to rays in a

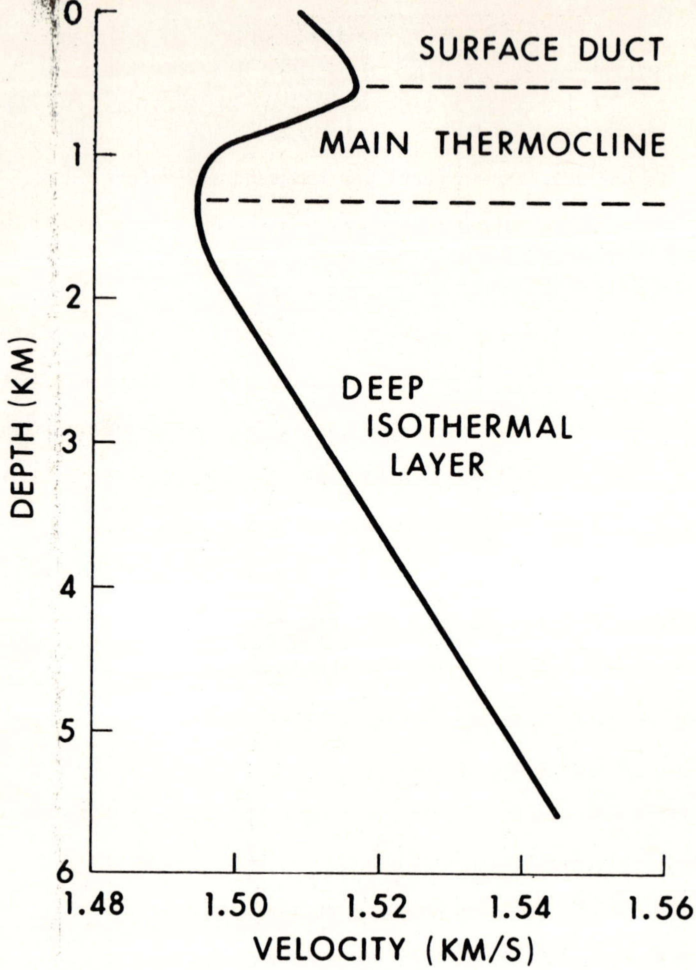

Figure 4. A typical deep ocean velocity-depth profile. The surface duct and deep isothermal layer are separated by the main thermocline.

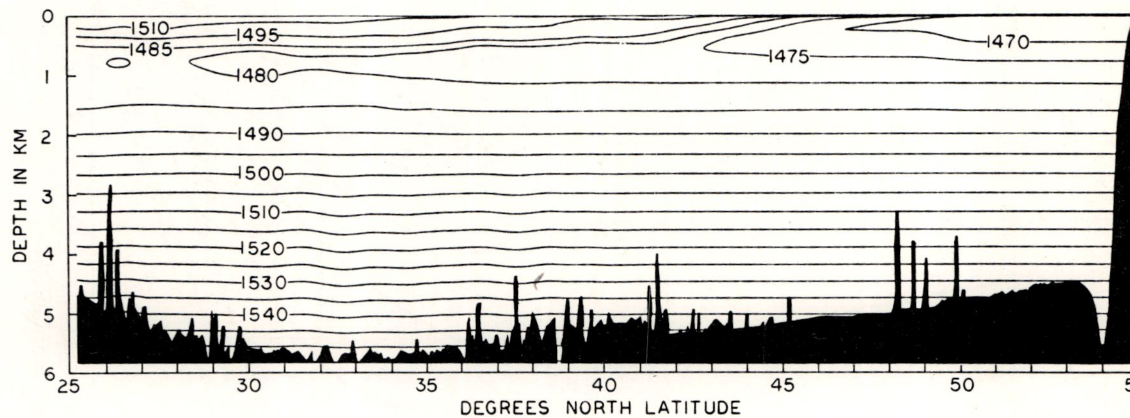

Figure 5. Contour of the sound speed measured during the late summer of 1968 along the meridian 157°50'W.

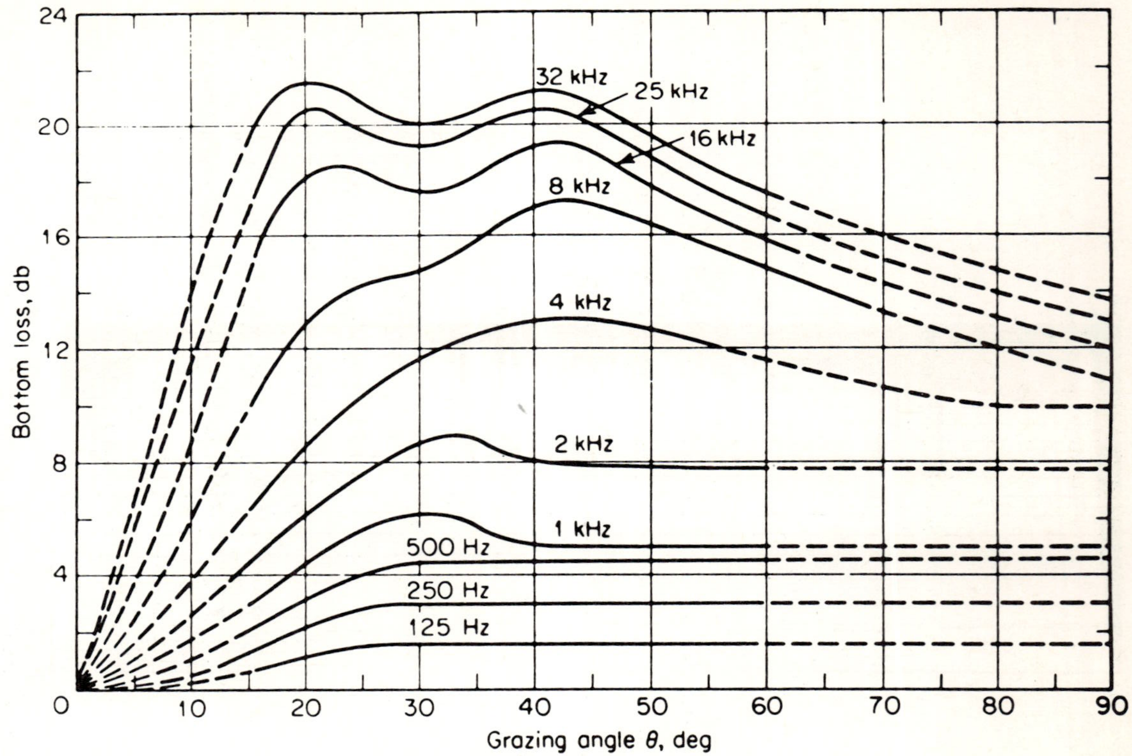

Figure 6. Bottom loss versus grazing angle at selected frequencies. The dashed lines indicate regions of little or no data. With permission of Urick [26].

vertical plane but a normal mode with no turning point can be associated in a natural way with such rays and will have a definite grazing angle associated with it. We see that larger grazing angles are associated with larger losses.

In the water itself the predominant absorption mechanisms are viscosity and ionic relaxation of $MgSO_4$. The most widely used formula for the attenuation coefficient is due to Thorp [27]

6.3 $$\alpha = \frac{1}{10}\frac{f^2}{1+f^2} + \frac{40f^2}{4100+f^2} ,$$

where f is frequency in kilohertz and α is the attenuation in decibels per kiloyard.

Although (6.3) is a good fit to measured data ([28]) above a few hundred hertz, a second relaxation phenomenon introduces discrepancies at lower frequencies. The equation

6.4 $$\alpha = 5.42\times10^{-2}\ f^{3/2}\ \mathrm{dB/kyd} ,$$

developed by Sussman, MacDonald and Kanabis [29] supplements the above results and should be used below a transitional frequency of about 280 Hz .

It follows from all these data that if we limit the frequency to around 100 Hz and wind speed to around 20 knots and consider horizontal ranges to beyond 10^5 yds. (50 to 60 miles) then we may assume that the ocean surface is free and we may neglect modes associated with large grazing angles.

6.2 The computer program

The computer program which we have developed to implement our scheme consists of two parts. The first determines the normalized eigenfunctions and eigenvalues at each point of a rectangular grid in the horizontal plane. In the second part the horizontal ray-tracing equations are integrated, and the contributions of individual modes are combined to obtain the total field. The propagation loss along any desired linear track in the horizontal plane may be displayed. Propagation loss is $-20 \log_{10}|\Phi/\Phi_o|$ where Φ is the acoustic pressure at a receiver on the track and Φ_o is the acoustic pressure 1 yd. away from the source. According to the law of reciprocity, one may also think of the source as being situated on the horizontal track and the receiver as being the origin of the horizontal rays.

The first part of the program requires as input the velocity-depth profile z_ℓ^{ij} , c_ℓ^{ij} , $\ell = 1,2,\ldots,n_{ij}$, at each horizontal lattice point. The lattice points are (x_i, y_j) and for each ij , c_ℓ^{ij} is the sound speed at depth z_ℓ^{ij}.

Temporarily dropping the superscripts ij we define

$$6.5 \qquad k_\ell = \frac{\omega}{c_\ell}$$

where ω is the frequency in radians per second. The derivative g_ℓ is defined by

$$6.6 \qquad g_\ell = \frac{k^2_{\ell+1} - k_\ell^2}{z_{\ell+1} - z_\ell}$$

and $k^2(z)$ is given by linear interpolation in the interval $(z_\ell, z_{\ell+1})$ as

$$6.7 \qquad k^2(z) = k_\ell^2 + g_\ell(z-z_\ell) .$$

It was found convenient to treat the refracted-surface-reflected (RSR) and the trapped (SOFAR) modes differently from the bottom-bounce (B) modes. Assuming that the eigenfunction corresponding to an RSR or SOFAR mode decays exponentially with depth in the deep isothermal layer $z_{n-1} \leq z \leq z_n$ we must have

$$6.8 \qquad C^{-1}\,\psi_\lambda(z) = d_{n-1}\mathrm{Ai}\,\{-g_{n-1}^{-2/3}[k_{n-1}^2 - g_{n-1}\,(z-z_{n-1}) - \lambda^2]\} ,$$

where C is a normalization constant. If $\ell < n - 1$

$$6.9 \qquad C^{-1}\,\psi_\lambda(z) = d_\ell\mathrm{Ai}\{-g_\ell^{-2/3}[k_\ell^2 + g_\ell(z-z_\ell) - \lambda^2]\}$$

$$+ e_\ell\mathrm{Bi}\{-g_\ell^{-2/3}[k_\ell^2 + g_\ell(z-z_\ell) - \lambda^2]\} ,$$

and

$$6.10 \qquad C^{-1} \frac{\partial\psi_\lambda(z)}{\partial z} = d_\ell\, g_\ell^{-1/3}\, Ai'\{-g_\ell^{-2/3}[k_\ell^2 + g_\ell(z-z_\ell) - \lambda^2]\}$$

$$+ e_\ell^{-1/3}\, Bi'\, \{-g_\ell^{-2/3}[k_\ell^2 + g_\ell(z-z_\ell) - \lambda^2]\} \; .$$

The constants d_ℓ , e_ℓ are determined successively by the continuity of ψ and $\partial\psi/\partial z$ at z_ℓ . The eigenvalues are found by shooting for $\psi_\lambda(0) = 0$. Once the λ are determined, Gaussian quadrature is used to compute the normalization constant C .

For the bottom-bounce modes we proceed from z = 0 by first setting

$$6.11 \qquad C^{-1}\, \psi_\lambda\, (0) = 0 \; ,$$

$$6.12 \qquad C^{-1} \frac{\partial\psi_\lambda(Z)}{\partial z} = 1 \; ,$$

successively evaluating the constants d_ℓ , e_ℓ for $\ell = 1,2,\ldots,n-1$ and then adjusting λ so that

$$6.13 \qquad C^{-1} \frac{\partial\psi_\lambda(z_n)}{\partial z} = 0 \; .$$

However, it is more nearly correct to require

$$6.14 \qquad \frac{\partial\psi_\lambda\,(z_n)}{\partial z} \Big/ \psi_\lambda(z_n) = a(\lambda^2) \; ,$$

where $a(\lambda^2)$ is determined from the empirical bottom-loss expressions. Since $a(\lambda^2)$ is small the eigenvalues for the B modes are modified accordingly to $\lambda_m'^2$, where

$$6.15 \qquad \lambda_m'^2 = \lambda_m^2 + a(\lambda_m^2) \left[\frac{\partial}{\partial(\lambda^2)} \left(\frac{\partial\psi_\lambda^1(z_n)}{\partial z} \Big/ \psi_\lambda\,(z_n) \right) \right]_{\lambda=\lambda_m} .$$

The advantage of this procedure over that of computing λ_m' directly is that lengthy complex arithmetic may thus be avoided. The imaginary part of λ_m is the predominant attenuation factor at long ranges and should not be neglected. On the other hand the program does neglect the imaginary part of the eigenfunction.

A significant reduction in computation could be achieved by using the Bohr-Sommerfeld and WBK approximations for the eigenvalues and eigenfunctions. However in this prototype of the program we have not used these.

After all horizontal lattice points have been treated, control is transferred to part two of the program where a family of horizontal rays is traced for each mode.

The fact that we are now allowing the eigenvalues to have non-zero imaginary parts introduces an extra difficulty. Since only those modes which correspond to eigenvalues with very small imaginary parts will propagate to large distances, we assume that

6.16 $$\lambda^2 = \lambda_{re}^2 + i\delta\lambda_{im}^2 ,$$

where δ is a small parameter. On writing the phase function θ in the form

6.17 $$\theta = \theta_{re} + i\delta\theta_{im} + 0(\delta^2)$$

and equating coefficients of δ^0 , δ^1 in the eikonal equation we obtain

6.18 $$(\nabla\theta_{re})^2 = \lambda_{re}^2 ,$$

6.19 $$2\,\nabla\theta_{re}\cdot\nabla\theta_{im} = \lambda_{im}^2 .$$

In order to work with λ_{re}^2 directly rather than λ_{re} we introduce a new variable

6.20 $$u = \int_0^s \lambda_{re}\, ds' ,$$

where s is arc length. In terms of u the ray tracing equations become

6.21 $$\frac{d^2x}{du^2} = 1/2 \frac{\partial\lambda_{re}^2}{\partial x} ,$$

6.22 $$\frac{d^2y}{du^2} = 1/2 \frac{\partial\lambda_{re}^2}{\partial y} ,$$

6.23 $$\frac{d\theta_{re}}{du} = \lambda_{re}^2 ,$$

6.24 $$\frac{d\theta_{im}}{du} = 1/2 \lambda_{im}^2 .$$

These equations are integrated by a predictor-corrector method. Since values of λ_{re}^2 and λ_{im}^2 are required everywhere we interpolate to find the values of these quantities between lattice points. Fitting λ_{re}^2 rather than λ_{re} gives more nearly correct behavior near $\lambda_{re}^2 = 0$.

Now let $\{x_j\}$ be a set of values of x , not necessarily coinciding with corresponding coordinates of the lattice points. Whenever a ray crosses a line $x = x_j$ the pertinent data are stored for future reference. The horizontal phase and amplitude for a particular mode at an arbitrary point is obtained later by interpolating between the stored data. The pressure amplitude is computed by summing over all propagating modes.

6.3 Comparison of computed amplitudes with observational data

Two assumptions underlying the design of the computer program are that the speed of sound in the ocean is known reasonably accurately and that the acoustic pressure variations are due to a point source varying harmonically in time.

However, since amplitude, or equivalently propagation loss data of the type in which we are interested are in fact gathered over a period of several days, the sound speed may change somewhat from its measured values during the course of the experiment. Furthermore, source levels must be high enough for the signal to be detected at large ranges. This is usually accomplished by using a dynamite explosion as source. The response at a particular frequency is then obtained by passing the received signal through a filter having finite bandwidth. In view of the above remarks we cannot expect predictions and calculations to be in exact agreement.

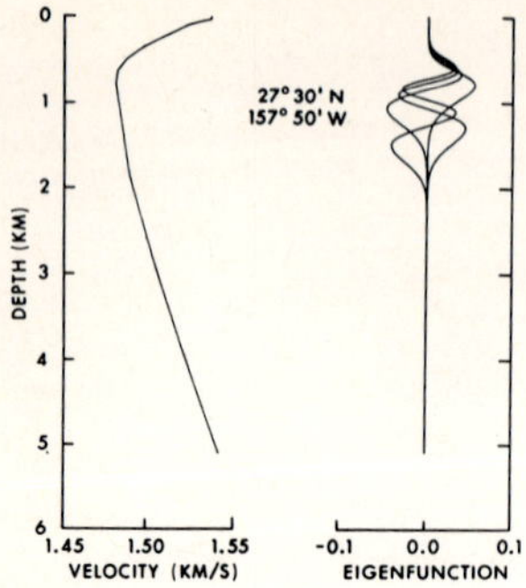
27° 30' N
157° 50' W
DEPTH (KM)
0
1
2
3
4
5
6
1.45
1.50
1.55
VELOCITY (KM/S)
-0.1
0.0
0.1
EIGENFUNCTION

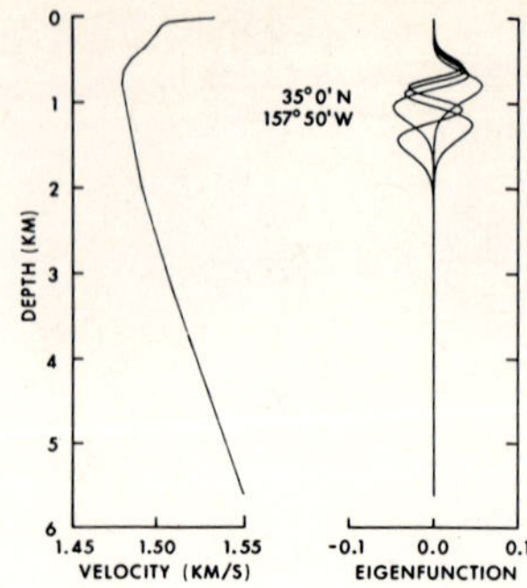
35° 0' N
157° 50' W
DEPTH (KM)
0
1
2
3
4
5
6
1.45
1.50
1.55
VELOCITY (KM/S)
-0.1
0.0
0.1
EIGENFUNCTION

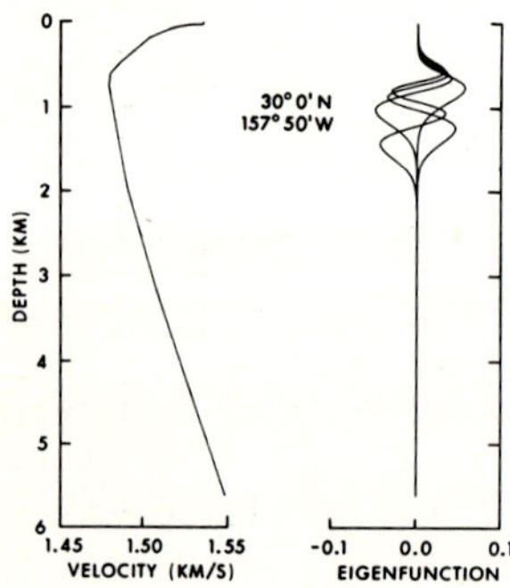
30° 0' N
157° 50' W
DEPTH (KM)
0
1
2
3
4
5
6
1.45
1.50
1.55
VELOCITY (KM/S)
-0.1
0.0
0.1
EIGENFUNCTION

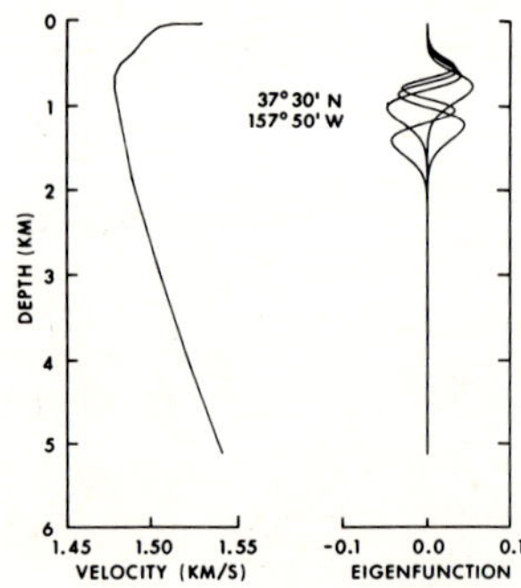
37° 30' N
157° 50' W
DEPTH (KM)
0
1
2
3
4
5
6
1.45
1.50
1.55
VELOCITY (KM/S)
-0.1
0.0
0.1
EIGENFUNCTION

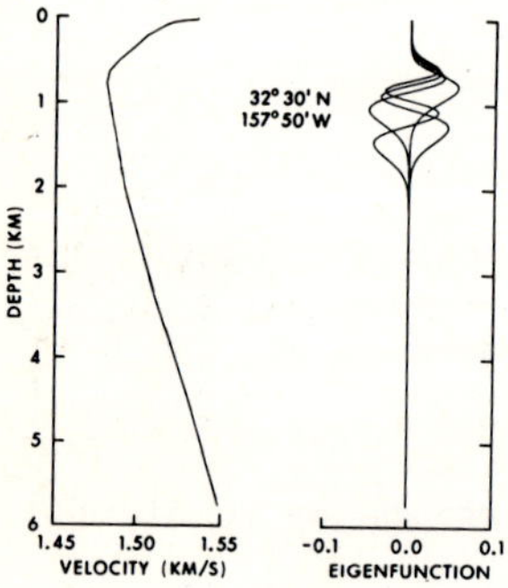
32° 30' N
157° 50' W
DEPTH (KM)
0
1
2
3
4
5
6
1.45
1.50
1.55
VELOCITY (KM/S)
-0.1
0.0
0.1
EIGENFUNCTION

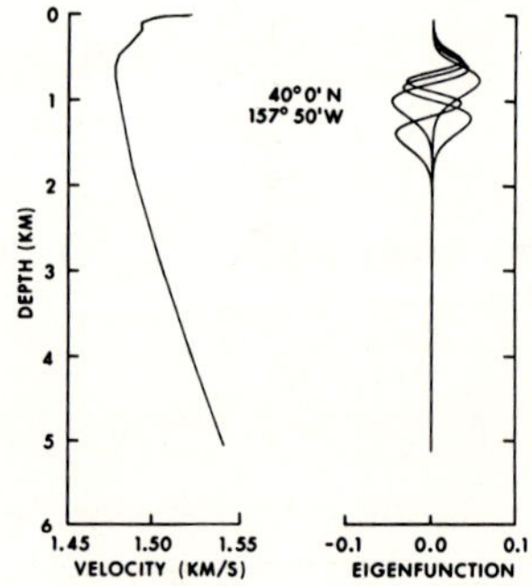
40° 0' N
157° 50' W
DEPTH (KM)
0
1
2
3
4
5
6
1.45
1.50
1.55
VELOCITY (KM/S)
-0.1
0.0
0.1
EIGENFUNCTION

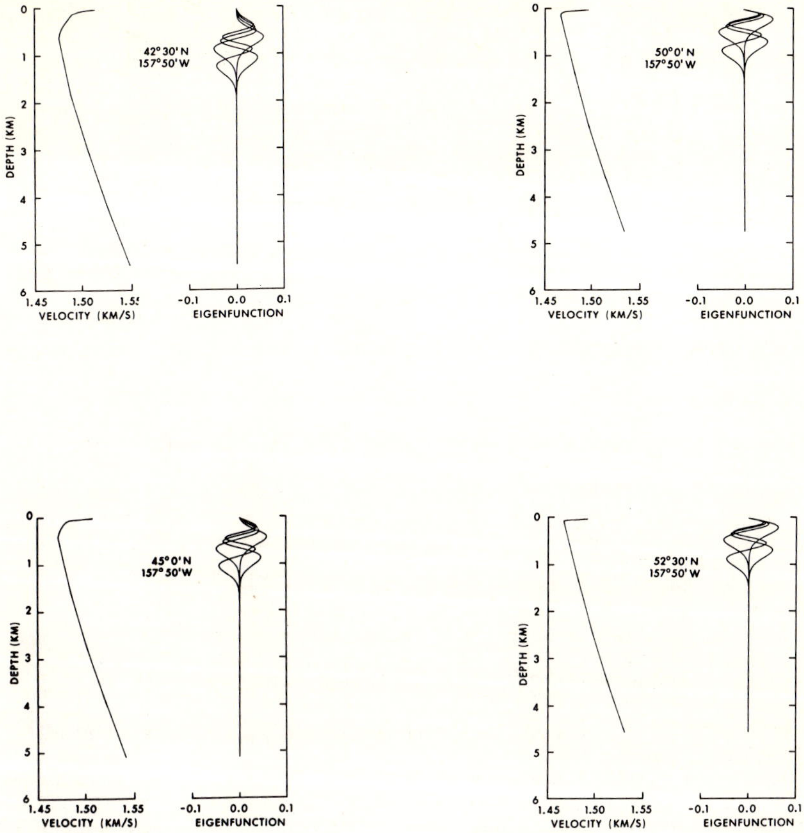

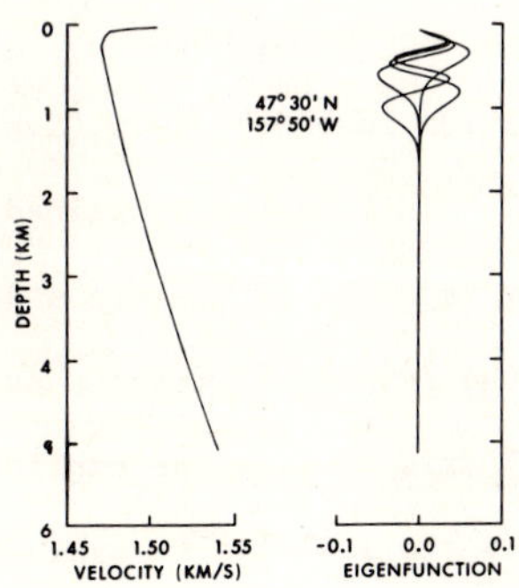

Figure 7. Velocity-depth profiles (left) and corresponding first four modes (right) at various geographical locations for a frequency of 31 hz.

In the particular experiment to be analyzed, dynamite charges were detonated 500 feet below sea level along a track 1500 nautical miles long, extending northward from 27°30' N, 157°50' W to 52°30' N, 157°50' W. Eleven equidistant velocity-depth profiles obtained from the measured data displayed in figure 5 were entered into the computer program. They are shown on the left in each frame of figure 7. We see that surface ducts were practically non-existent. Lack of relevant data prevented us from including any dependence of sound speed or bottom depth upon longitude.

The computer program was directed to determine 100 modes for each velocity-depth profile using a frequency of 31 hz. The first four modes for each profile are illustrated on the right in each frame of figure 7. Note that the fundamental modes are centered about the SOFAR axis, which rises from a depth of 2608 ft. (795 meters) at 27°30' N to about 164 feet (50 meters) at 52°30' N. On the average, for each profile, 45 modes corresponded to RSR or SOFAR modes while the remaining 55 were B modes. Figures 8 and 9 display propagation losses for receivers at depths of 2500 ft. and 10,800 ft. respectively, situated at 27°30' N, 157°50' W. The top graph in each figure represents observational data while the middle graph shows computed results. The measured data and the computer predictions are superimposed in the bottom graph. Peak values of measurements and predictions agree to within a few decibels along the entire 1500 nautical miles of the track. The computer program did predict nulls of about 10 db. in magnitude which were not found in the data. Such sharp minima, if they really occurred, could be missed owing to the finite spacing of the source points, or, fluctuations in the ocean and the receiving filter could smooth them out.

Figure 8 displays an interesting feature. The propagation loss decreases (i.e. the amplitude increases) with increasing range beyond 42° N. This may be explained by the fact that the 2500 ft. receiver is only 124 ft. away from the SOFAR axis and the signal there is strongly affected by the amplitude of the few lowest modes. As the source ship moved north the source approached the SOFAR axis causing the amplitude of these modes to increase to such an extent that eventually the loss due to cylindrical spreading was overcome and the total propagation loss decreased.

The 10,800 ft. receiver, on the other hand, is well below the turning points of the first few modes and so the signal there is dominated by the higher modes. The

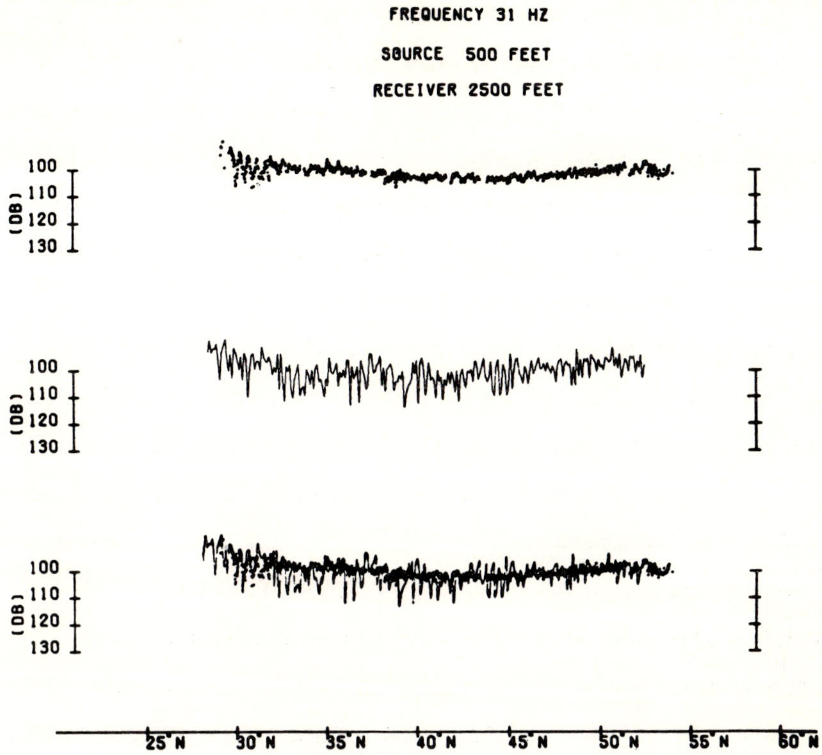

Figure 8. Propagation loss versus range for a receiver at a depth of 2500 ft., a source at 500 ft., and a frequency of 31 hz. The receiver is fixed at 27°30' N, 157°50' W while the source moves northward. The top, central, and lower curves represent measured data, computer predictions, and measurements superimposed on predictions, respectively.

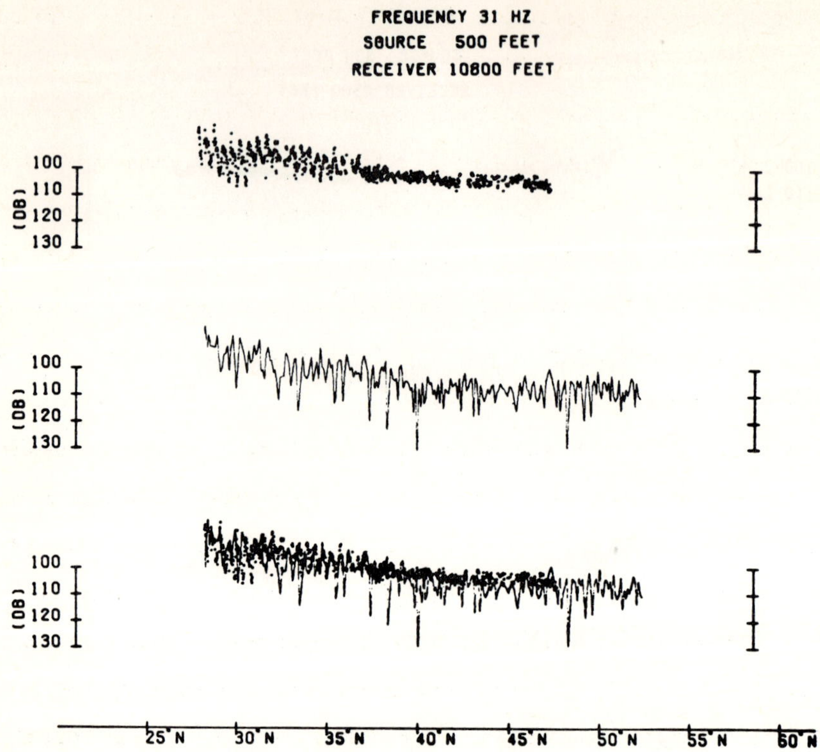

Figure 9. Same as Figure 8 except that the receiver is at a depth of 10,800 ft.

amplitudes of these are not greatly affected when the source approaches the SOFAR axis so that for this receiver cylindrical spreading dominates over the entire track.

We should remark here that originally our computations differed systematically from the observed data by a few decibels. After thoroughly checking the calculation and finding no error we were led to question the data. It turned out that as the data were compiled the equivalent source strength at 31 hz of the dynamite charges had been systematically overestimated. When the original source strength was replaced by the most recent estimate available to us we obtained the good agreement displayed in the bottom graphs in figures 7 and 8.

A limited number of cases were also investigated where certain parameters were varied to see what effect, if any, variations in surface loss, bottom loss, and attenuation would have on the above results. It was found that bottom loss had no detectable effect whatsoever, and that the results depended only weakly upon the attenuation. It was also determined that if the ocean had had East-West gradients in sound speed comparable with those in the North-South direction the radius of curvature of the rays corresponding to the dominant modes would be no less than $2. \times 10^5$ nautical miles. This implies firstly that negligible error was committed in neglecting the dependence on longitude, and secondly that in this particular problem the ray tracing procedure could have been replaced by the simple cylindrical spreading law.

References

[1] C. L. Pekeris, Theory of propagation of explosive sound in shallow water, Geol. Soc. Amer. Mem. 27 (1948).

[2] J. L. Worzel and M. Ewing, Explosive sounds in shallow water, Geol. Soc. Amer. Mem. 27 (1948).

[3] I. Tolstoy, Shallow water test of the theory of layered wave guides, J. Acoust. Soc. Amer. 30 384 (1958).

[4] A. D. Pierce, Extension of the method of normal modes to sound propagation in an almost stratified medium, J. Acoust. Soc. Amer. 37 (1965).

[5] J. B. Keller, Surface waves on water of nonuniform depth, J. Fluid Mech. 4 607 (1958).

[6] M. C. Shen and J. B. Keller, Ray method for nonlinear wave propagation in a rotating fluid of variable depth, Phys. Fluids 16 1565 (1973).

[7] B. Rulf, An asymptotic theory of guided waves, J. Engr. Math. 4 261 (1970).

[8] F. P. Bretherton, Propagation in slowly varying waveguides, Proc. Roy. Soc. A302 555 (1968).

[9] J. B. Keller, R. M. Lewis and B. D. Seckler, Asymptotic solutions of some diffraction problems, Comm. Pure Appl. Math. 9 207 (1956).

[10] F. C. Karal and J. B. Keller, Elastic wave propagation in homogeneous and inhomogeneous media, J. Acoust. Soc. Amer. 31 694 (1959).

[11] H. Weinberg and R. Burridge, Horizontal ray theory for ocean acoustics. J. Acoust. Soc. Amer. 55 63 (1974).

[12] D. Ludwig, Uniform asymptotic expansions at a caustic, Comm. Pure Appl. Math. 19 215 (1966).

[13] V. M. Babich, The short wave asymptotic form of the solution for the problem of a point source in an inhomogeneous medium, Z. Vycisl. Mat. i Mat. Fiz 5 969 (1965).

[14] E. Zauderer, Uniform asymptotic solutions of the reduced wave equation, J. Math. Anal. Appl. 30 157 (1970).

[15] J. Hadamard, Lectures on Cauchy's Problem in Linear Partial Differential Equations, Dover Publications, New York, (1952).

[16] P. R. Garabedian, Partial Differential Equations, Wiley, New York (1964), p. 27.

[17] W. Magnus and F. Oberhettinger, Formulas and Theorems for the Functions of Mathematical Physics, Chelsea, New York (1954) p. 22.

[18] H. Jeffreys and B. S. Jeffreys, Methods of Mathematical Physics, Cambridge University Press (1956), p. 511.

[19] F. Tappert, The parabolic equation method in underwater acoustics, This volume.

[20] E. Zauderer, On a modification of Hadamard's method for obtaining fundamental solutions for hyperbolic and parabolic equations, J. Inst. Math. Applics. 8 8 (1971).

[21] R. A. Handelsman and N. Bleistein, Uniform asymptotic expansions of integrals that arise in the analysis of precursors, Arch. Rat. Mech. Anal. 35 267 (1969).

[22] V. Cerveny and R. Ravindra, Theory of Seismic Head Waves, University of Toronto Press (1971).

[23] C. C. Leroy, Development of simple equations for accurate and more realistic calculation of the speed of sound in sea water, J. Acoust. Soc. Amer. 46 216 (1969).

[24] R. H. Adlington, Acoustic reflection losses at the sea surface measured with explosive sources, J. Acoust. Soc. Amer. 35 1834 (1963).

[25] H. W. Marsh, T. G. Bell, and C. W. Horton, Reflection and scattering of sound by the sea bottom, in Proceedings of the 68th meeting of Acoust. Soc. Amer. Oct. 21-24, 1964, Austin, Texas (Jan. 1965).

[26] R. J. Urick, Principles of Underwater Sound for Engineers (McGraw-Hill Book Company, New York, 1967), p. 118.

[27] W. H. Thorp, Analytic description of the low-frequency attenuation coefficient, Letter to Editor, J. Acoust. Soc. Amer. 42 270 (1967).

[28] W. H. Thorp, Deep ocean sound attenuation in the sub- and low-kilocycle-per-second region, J. Acoust. Soc. Amer. 38 648 (1965).

[29] R. D. Sussman, R. D. MacDonald, and W. G. Kanabis, Frequency-attenuation studies, Under-Water Sound Laboratory Tech. Memo. No. 911-16-65 (Oct. 1964).

CHAPTER IV

WAVE PROPAGATION IN A RANDOMLY INHOMOGENEOUS OCEAN

Werner Kohler

Virginia Polytechnic Institute
and State University
Blacksbourg, VA 24061

George C. Papanicolaou

Courant Institute of Mathematical Sciences
New York University
251 Mercer Street
New York, NY 10012

0. Introduction

The purpose of this chapter is to present in a self-contained manner an analysis of some phenomena associated with random fluctuations of the sound speed of the ocean.

In section 1 we formulate the underwater sound problem in a manner convenient for the stochastic analysis. We introduce several simplifying assumptions, such as the forward scattering approximation, but we maintain radiation losses into the ocean bottom. We employ a modal decomposition relative to the modes of the mean soundspeed profile. The resulting set of stochastic equations for the mode amplitudes as functions of range is the starting point of the stochastic analysis.

In section 2 we give a brief but self-contained description of the relevant asymptotics for stochastic equations. The procedure described is nothing more than second order perturbation theory applied properly. More information regarding stochastic problems can be found in [5-12]. References [8],[9] and [11] contain many interesting examples and introduce the methodology systematically while keeping mathematics at a formal level. More mathematical treatments are [12], [17] and [21].

In section 3 we apply the method outlined in section 2 to the underwater sound problem. The result contains, in principle, the complete probabilistic characterization of the complex-valued mode amplitudes in the relevant asymptotic limit. This limit corresponds to propagation over distances that are large compared to the horizontal correlation length of the soundspeed inhomogeneities, and to weak fluctuations in the soundspeed from its mean value. The wavelength is assumed to be of order one relative to the correlation length.

Sections 4-8 contain concrete information about the underwater sound problem that can be obtained by specializing the results of section 3.

In section 4 we derive the coupled power equations. They control the dynamics (as functions of range) of the mean power transfer between the trapped (or propagating) modes and radiation losses. We feel, as does Marcuse [16], for example, for the corresponding optical fiber problem, that the coupled power equations should be an important tool in analyzing fluctuation phenomena. We illustrate this in section 5 where we take up the pulse spreading (in time) problem and show how to obtain Personick's results [18,19] in the present context.

In section 6 we derive equations for the evolution with range of the fluctuations in modal powers about their mean values. These equations lead to some interesting conclusions when the number of trapped modes is large. There are many interesting problems in connection with power fluctuations that have not been analyzed yet. One can find some conjectures in [16], for example. Of course, one can also study higher moments and the statistics of relative phases of the mode amplitudes. The set up of section 3 contains all this information but it is a major task (possibly numerical) to extract it from there without additional simplifying assumptions.

In section 7 we indicate very briefly how to calculate statistics of depth dependent quantities by superposing modes and using results of previous sections.

In section 8 we examine the form of the coupled power equations at high frequency i.e., when the number of propagating modes is large. We find that they are well approximated by a diffusion equation where range plays the role of time and ray angle plays the role of space variable. Such diffusion equations have been obtained before [23,24] by physical arguments that seem quite natural (cf. also [25],[26]). One can also give a derivation of these diffusion equations directly without first going to the coupled power equations (we do not do this here). It would be interesting to obtain comparable results for the coupled fluctuation equations. Numerical comparisons show that the diffusion equation is a very good approximation to the coupled power equations even when the number of propagating modes is not too large (say 10-20). This is another reason why a diffusion approximation for the coupled fluctuation equations would be very useful.

We wish to thank L. Dozier for reading the manuscript and suggesting several improvements (cf. also [27] for some interesting results extending some of the analysis given here).

1. The physical problem

Let $p(r,\theta,z)$ denote the sound pressure field in cylindrical coordinates with z measured downward from the surface of the ocean (figure 1) and with the time factor $e^{-i\omega t}$ omitted throughout. The pressure satisfies the following equation and boundary conditions:

1.1
$$\frac{\partial^2 p}{\partial r^2} + \frac{1}{r}\frac{\partial p}{\partial r} + \frac{1}{r^2}\frac{\partial^2 p}{\partial \theta^2} + \frac{\partial^2 p}{\partial z^2} + k^2[n^2(z)+\varepsilon\mu(r,z)]p = \frac{\delta(r)}{2\pi r}\,\delta(z-z_o)\ ,$$

$$r \geq 0, \quad 0 \leq \theta < 2\pi, \quad 0 \leq z < \infty, \quad p(r,\theta,0) = 0\ .$$

Here $n(z)$ denotes the mean index of refraction, $n(z) = c_o/c(z)$ where $c(z)$ is the mean velocity profile. This mean index of refraction is assumed to be a function of depth only. The fluctuations about the mean are denoted by $\mu(r,z)$; they are random and they can vary with range and depth. We have assumed that the fluctuations do not depend on the azimuthal angle in order to simplify the analysis that follows. The

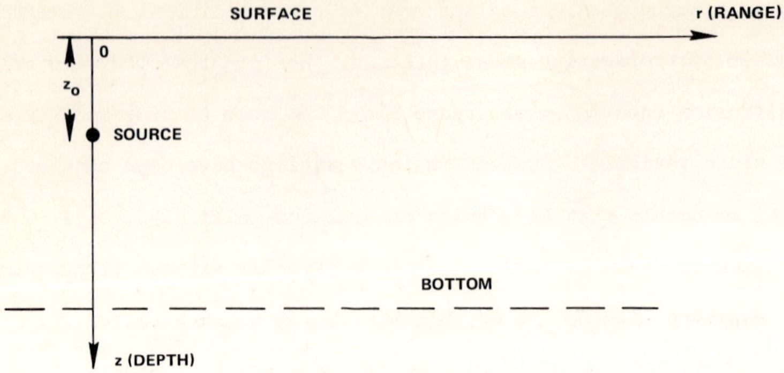

Figure 1

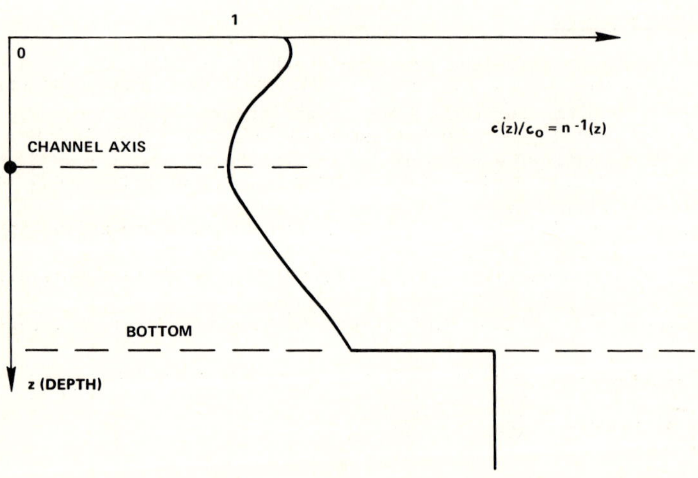

Figure 2

general case entails no essential difficulties.

The parameter ε in (1.1) characterizes the size of the fluctuations and it is typically small, $\varepsilon \sim 10^{-2}$. The fluctuations by definition have mean zero

$$<\mu(r,z)> = 0 \ , \tag{1.2}$$

where $<\cdot>$ denotes ensemble average or expectation value. We also assume that $\mu(r,z) \equiv 0$ for z sufficiently large, i.e., inside the ocean floor.

In view of the azimuthal symmetry, $\frac{1}{2\pi}\int_0^{2\pi} p(r,\theta,z)d\theta$, also denoted by $p(r,z)$, satisfies the simpler equation

$$\frac{\partial^2 p}{\partial r^2} + \frac{1}{r}\frac{\partial p}{\partial r} + \frac{\partial^2 p}{\partial z^2} + k^2[n^2(z)+\varepsilon\mu(r,z)]p = \frac{\delta(r)}{2\pi r}\,\delta(z-z_o) \ , \tag{1.3}$$

$$0 \le r < \infty, \quad 0 \le z < \infty, \quad p(r,0) = 0 \ ,$$

corresponding to (1.1).

Next, we introduce the assumption that the stochastic effects we seek to analyze manifest themselves entirely within the cylindrically spreading regime. This means that there is a region around the source location large enough so that the emitted spherical waves have reached their asymptotic cylindrically spreading state. At the same time this region is small enough so that stochastic effects have not accumulated and can be ignored. The precise analysis of matching the field of a point source to the cylindrically spreading regime is given in [1] . We assume here that we may replace p by $p/\sqrt{r}$; symbolically

$$p(r,z) \to \frac{p(r,z)}{\sqrt{r}} \ , \tag{1.4}$$

so the new p is scaled by the geometrical spreading factor. Neglecting a near field term of the form $p/r^{5/2}$ the scaled pressure field satisfies

$$\frac{\partial^2 p}{\partial r^2} + \frac{\partial^2 p}{\partial z^2} + k^2[n^2(z)+\varepsilon\mu(r,z)]p = 0 \ , \tag{1.5}$$

$$r > 0, \quad z \geq 0, \quad p(r,0) = 0 .$$

In (1.5) the values of p are prescribed in some way at $r = 0$ by the matching-to-the-source procedure. Since we shall also employ a forward scattering approximation we postpone discussion of this until later.

Let us consider the differential operator

$$1.6 \qquad L \equiv \frac{d^2}{dz^2} + k^2 n^2(z), \quad z > 0 ,$$

zero boundary condition at $z = 0$.

This operator is selfadjoint in $L^2(0,\infty)$ for a broad class of (normalized) indices of refraction of the form shown in figure 2 ([3]). Its spectrum contains finitely many discrete eigenvalues and a semi-infinite line, the continuous or radiation spectrum. We assume that the eigenvalues and eigenfunctions (the modes) satisfy the following equations and normalization conditions.

$$1.7 \qquad L\nu_p(z) = \beta_p^2 \nu_p(z), \quad \nu_p(0) = 0, \quad p = 1,2,\ldots,N$$

$$1.8 \qquad L\nu(z,\gamma) = \gamma\nu(z,\gamma), \quad \nu(0,\gamma) = 0, \quad -\infty \leq \gamma \leq k^2$$

$$1.9 \qquad (\nu_p,\nu_q) = \int_0^\infty \nu_p(z)\nu_q(z)dz = \delta_{pq}$$

$$1.10 \qquad (\nu_p,\nu(\gamma)) = 0, \quad (\nu(\gamma),\nu(\gamma')) = \delta(\gamma-\gamma') ,$$

where,

$$1.11 \qquad \nu_p = \nu_p(\cdot), \quad \nu(\gamma) = \nu(\cdot,\gamma)$$

and we have assumed that the mode functions are real. Note that $N = N(k)$ i.e., the number of discrete or trapped modes depends on the wave number $k = \omega/c_o$ and it increases as k increases. The eigenvalues $\beta_p = \beta_p(k)$ are the propagation constants of the trapped modes. The ocean is evidently a dispersive medium since modes travel at different speeds $c_o[\partial\beta_p(k)/\partial k]^{-1}$.

Let us expand the solution p(r,z) of (1.5) in terms of the eigenfunctions of L ,

1.12 $$p(r,z) = \sum_{p=1}^{N} c_p(r)\nu_p(z) + \int_{-\infty}^{k^2} c(\gamma,r)\nu(z,\gamma)d\gamma \ .$$

On inserting this expression into (1.5) and using the orthonormality conditions (1.9), (1.10), we obtain the following equations for the <u>mode amplitudes</u> $c_p(r)$ and $c(\gamma,r)$.

1.13 $$\frac{d^2c_p(r)}{dr^2} + \beta_p^2 c_p(r) + \varepsilon k^2 \sum_{q=1}^{N} \hat{\mu}_{pq}(r)c_q(r)$$
$$+ \varepsilon k^2 \int_{-\infty}^{k^2} \hat{\mu}_{p\gamma}(r)c(\gamma,r)d\gamma = 0 \ , \qquad p = 1,2,\ldots,N \ ,$$

1.14 $$\frac{d^2c(\gamma,r)}{dr^2} + \gamma c(\gamma,r) + \varepsilon k^2 \sum_{q=1}^{N} \hat{\mu}_{\gamma q}(r)c_q(r)$$
$$+ \varepsilon k^2 \int_{-\infty}^{k^2} \hat{\mu}_{\gamma\gamma'}(r)c(\gamma',r)d\gamma' = 0 \ , \qquad -\infty < \gamma \leq k^2 \ .$$

In (1.13) and (1.14) we have denoted by $\hat{\mu}$ with subscripts the matrix elements of $\mu(r) = \mu(r,\cdot)$, the fluctuation function, with respect to the modes:

1.15 $$\hat{\mu}_{pq}(r) = (\mu(r)\nu_q,\nu_p), \qquad \hat{\mu}_{p\gamma}(r) = (\mu(r)\nu(\gamma),\nu_p) \ ,$$
$$\hat{\mu}_{\gamma q}(r) = (\mu(r)\nu_q,\nu(\gamma)), \quad \hat{\mu}_{\gamma\gamma'}(r) = (\mu(r)\nu(\gamma'),\nu(\gamma)) \ .$$

In the way these matrix elements are arranged in (1.15) they form a real symmetric "matrix", the quotation indicating that some entries in the matrix are continuously indexed (the subscripts γ and γ' range over $-\infty < \gamma,\gamma' \leq k^2$) .

We shall assume in the following that the evanescent continuous modes $\nu(\gamma,z)$,

$-\infty < \gamma \leq 0$ can be neglected. This is a reasonable assumption because these waves do not propagate energy over long distances. It is also an assumption compatible with the forward scattering (or parabolic) approximation which we introduce next.

Let us write the mode amplitudes in the following form[†]

$$1.16 \qquad c_p(r) = \frac{1}{\sqrt{\beta_p}} [c_p^+(r) e^{i\beta_p r} + c_p^-(r) e^{-i\beta_p r}] , \qquad p = 1,2,\ldots,N$$

$$c(\gamma,r) = \frac{1}{\gamma^{1/4}} [c^+(\gamma,r) e^{i\sqrt{\gamma} r} + c^-(\gamma,r) e^{-i\sqrt{\gamma} r}] , \qquad 0 < \gamma \leq k^2$$

The complex random functions $c_p^{\pm}(r)$ and $c^{\pm}(\gamma,r)$ are called the <u>forward</u>, with $+$, and the <u>backward</u>, with $-$, <u>propagation</u> or <u>mode amplitudes</u>. This is consistent with the assumed time factor $e^{-i\omega t}$. Since for each $p = 1,2,\ldots,N$ and for each $0 < \gamma \leq k^2$, a pair of complex functions is introduced, we may prescribe one additional relation for the pair. We take these to be

$$1.17 \qquad e^{i\beta_p r} \frac{dc_p^+(r)}{dr} + e^{-i\beta_p r} \frac{dc_p^-(r)}{dr} = 0$$

$$e^{i\sqrt{\gamma} r} \frac{dc^+(\gamma,r)}{dr} + e^{-i\sqrt{\gamma} r} \frac{dc^-(\gamma,r)}{dr} = 0 .$$

We insert next (1.16) into (1.13) and (1.14) and use (1.17). This way we obtain coupled equations for $c_p^{\pm}(r)$ and $c^{\pm}(\gamma,r)$ which involve only first order derivatives in r. We assume that $c_p^-(r)$ and $c^-(\gamma,r)$ can be neglected in these equations. This constitutes the <u>forward scattering approximation</u>. Its justification in the context of the stochastic problem rests mostly on the available evidence, experimental and numerical ([4]). It is believed to be very good for the underwater sound problem within a broad range of frequencies. At the end of Section 3 we give a more precise criterion, in terms of the statistical properies of the inhomogeneities and other quantities, that determines the range of validity of this assumption.

---

† The factors $\beta_p^{-1/2}$ and $\gamma^{-1/4}$ are introduced in order that the coefficients in (1.19) will be symmetric.

With the forward scattering approximation, the equations for the forward propagating complex mode amplitudes are as follows (the superscript + is omitted from now on).

$$1.18 \qquad \frac{dc_p(r)}{dr} = \varepsilon i \sum_{q=1}^{N} \mu_{pq}(r) e^{i(\beta_q - \beta_p)r} c_q(r) + \varepsilon i \int_0^{k^2} \mu_{p\gamma}(r) e^{i(\sqrt{\gamma} - \beta_p)r} c(\gamma, r) d\gamma \ ,$$

$$p = 1,2,\ldots,N$$

$$\frac{dc(\gamma,r)}{dr} = \varepsilon i \sum_{q=1}^{N} \mu_{\gamma q}(r) e^{i(\beta_q - \sqrt{\gamma})r} c_q(r) +$$

$$+ \varepsilon i \int_0^{k^2} \mu_{\gamma\gamma'}(r) e^{i(\sqrt{\gamma'} - \sqrt{\gamma})r} c(\gamma', r) d\gamma' \ , \qquad 0 < \gamma \leq k^2 \ .$$

Here we have introduced the notation

$$1.19 \qquad \mu_{pq}(r) = \frac{k^2}{2\sqrt{\beta_p \beta_q}} \hat{\mu}_{pq}(r), \qquad \mu_{p\gamma}(r) = \frac{k^2}{2\sqrt{\beta_p}\sqrt{\gamma}} \hat{\mu}_{p\gamma}(r) \ ,$$

$$\mu_{\gamma' p}(r) = \frac{k^2}{2\sqrt{\beta_p}\sqrt{\gamma'}} \hat{\mu}_{\gamma' p}(r), \qquad \mu_{\gamma'\gamma}(r) = \frac{k^2}{2(\gamma\gamma')^{1/4}} \hat{\mu}_{\gamma'\gamma}(r) \ .$$

We must now assign initial values at $r = 0$ for the system (1.18). This brings us back to the remark following (1.5) namely, that initial values for the pressure field (i.e. at zero range) must be obtained by matching the cylindrically spreading wave to a spherical wave. We shall assume that this has been done ([1]) and that

$$1.20 \qquad c_p(0) = c_{po}, \qquad p = 1,\ldots,N$$

$$c(\gamma,0) = c_o(\gamma), \qquad 0 < \gamma \leq k^2$$

where c_{po} and $c_o(\gamma)$ are given complex numbers. These numbers characterize the nature of the source, i.e., the manner in which the source transfers energy into the trapped and the radiation modes.

It is convenient to introduce matrix notation to represent the system (1.18). We must allow, however, for discrete as well as continuous indices since $p = 1,2,\dots,N$ and $0 < \gamma \leq k^2$. With this convention we write†

1.21 $$\mu(r) = \begin{bmatrix} \mu_{pq}(r) & \mu_{p\gamma}(r) \\ & \\ \mu_{\gamma' q}(r) & \mu_{\gamma\gamma'}(r) \end{bmatrix}$$

with the entries defined by (1.19). From our assumption $\mu(r)$ is a real-valued symmetric random matrix. If we also introduce the discrete and continuously indexed vector

1.22 $$c(r) = \begin{bmatrix} c_1(r) \\ \vdots \\ c_N(r) \\ \vdots \\ c(\gamma,r) \\ \vdots \end{bmatrix} , \qquad 0 < \gamma \leq k^2$$

and the diagonal matrix

1.23 $$\beta = \text{diag.}(\beta_1,\beta_2,\dots,\beta_N,\dots,\sqrt{\gamma},\dots) , \qquad 0 < \gamma \leq k^2 ,$$

then we may write (1.18) in the compact form

1.24 $$\frac{dc(r)}{dr} = \varepsilon i e^{-i\beta r}\mu(r)e^{i\beta r}\, c(r) , \quad r > 0 , \quad c(0) = c_o .$$

† The random function $\mu(r,z)$ in (1.1) will not be used in the sequel so the notation will cause no confusion.

Here $\mu(r)$ acts as a matrix in the discrete indices and as an integral operator in the continuous indices.

We shall refer to (1.24) as the <u>coupled mode equations</u>. They constitute an infinite system of coupled stochastic equations in slowly varying form, i.e., with ε multiplying the terms on the right hand side. The matrix of linear operators on the right side of (1.24) is random with mean zero, $\langle\mu(r)\rangle = 0$, in view of (1.2). In addition $\mu(r)$ is statistically stationary, a property that derives from the stationarity of the fluctuations $\mu(r,z)$ as random functions of the range. We do not assume stationarity in the depth variable. Since $\mu(r)$ is real and symmetric it follows from (1.24) that the total energy

$$\sum_{p=1}^{N} |c_p(r)|^2 + \int_0^{k^2} |c(\gamma,r)|^2 d\gamma \ , \tag{1.25}$$

carried by the forward propagating trapped and radiation modes is conserved, i.e., it is independent of the range r .

In the next section we discuss the methods for analysis of the stochastic equations (1.24). These methods are then applied to (1.24) in Section 3.

2. <u>Asymptotic analysis of stochastic equations</u>

We wish to analyze the behavior of the statistical properties of the solution $c(r)$ of (1.24), i.e., the statistics of the complex mode amplitudes as functions of the range r, when ε is small. In order to describe as simply as possible the essential points in the asymptotic analysis we shall restrict attention in this section to finite-dimensional systems of the form

$$\frac{dy_p(r)}{dr} = \varepsilon \sum_{q=1}^{N} V_{pq}(r) y_q(r) \ , \ r > 0 \ , \ \ y_p(0) = y_{po} \ , \ \ p = 1,2,\dots,N \tag{2.1}$$

$$V_{pq}(r) = ie^{i(\beta_q-\beta_p)r} \mu_{pq}(r) \ , \tag{2.2}$$

or, more compactly,

2.3 $$\frac{dy(r)}{dr} = \varepsilon V(r)y(r)\,, \qquad y(0) = y_o\,.$$

The matrix $\mu(r) = (\mu_{pq}(r))$ is assumed to have entries which are real, zero-mean, stationary random processes and to be symmetric. Note that (2.1) or (2.3) is a complex system which we could write as a real system of twice the complex dimension N. Again, for the purposes of this section we shall assume that (2.3) is a real system with V(r) a general real matrix valued process with zero-mean. We shall not assume however that V(r) is stationary since, in view of the exponential factors in (2.2), it is not in the example of interest to us. The presence of the oscillatory exponential factors plays an important role in the analysis that follows.

We are interested in the behavior of y(r) when ε is small but r is large so that cumulative fluctuation effects have had the opportunity to develop. Specifically, we shall allow r to vary in the interval $0 \leq r \leq \tau_o/\varepsilon^2$ where τ_o is some finite number which is arbitrary but fixed.[†] It is in this range that such stochastic effects emerge. We shall describe at first the behavior of $\langle y(\cdot)\rangle$, the expectation of $y(\cdot)$. This is no restriction in generality because (2.3) is generic in form. For example

$$\frac{d}{dr}(y_p(r)y_{p'}(r)) = \varepsilon \sum_{q,q'=1}^{N} [V_{pq}(r)\delta_{p'q'} + \delta_{pq}V_{p'q'}(r)]y_q(r)y_{q'}(r) \quad ,$$

which is again an equation of the form (2.3) for the doubly indexed vector $y_p(r)y_{p'}(r)$, $p,p' = 1,2,\ldots,N$. Later on we shall describe how one can obtain the behavior of the full probability distribution of $y(\cdot)$, which is our main objective here.

Let us rewrite (2.3) in integrated form

2.4 $$y(r) = y_o + \varepsilon \int_0^r V(s)y(s)ds\,.$$

† Frequently the results below hold with $\tau_o = \infty$. Then the assumption $\tau_o < \infty$ represents no essential loss in generality, and it is not necessary to specify τ_o numerically.

Upon iterating this equation once we obtain

$$2.5 \qquad y(r) = y_o + \varepsilon \int_0^r V(s)y_o ds + \varepsilon^2 \int_0^r\int_0^s V(s)V(\sigma)y(\sigma)d\sigma ds \ .$$

We now take ensemble averages in (2.5) and use the hypothesis $\langle V(r)\rangle = 0$. This yields

$$2.6 \qquad \langle y(r)\rangle = y_o + \varepsilon^2 \int_0^r \int_0^s \langle V(s)V(\sigma)y(\sigma)\rangle d\sigma ds \ ,$$

which appears to pose a "closure" problem since higher moments enter. Under certain hypotheses on $V(r)$, which we explain below, one can show that

$$2.7 \qquad \langle y(r)\rangle = y_o + \varepsilon^2 \int_0^r \int_0^s \langle V(s)V(\sigma)\rangle\langle y(\sigma)\rangle d\sigma ds + O(\varepsilon^3) \ , \quad 0 \le r \le \tau_o/\varepsilon^2 \ .$$

On dropping the $O(\varepsilon^3)$ on the right side of (2.7) one obtains the <u>first order smoothing approximation</u> to $\langle y(r)\rangle$ ([5-9]) which we shall continue to denote by $\langle y(r)\rangle$.

In order to arrive at results that are sufficiently simple and useful one must continue beyond the smoothing approximation. First we rewrite (2.7) in differential form

$$2.8 \qquad \frac{d\langle y(r)\rangle}{dr} = \varepsilon^2 \int_0^r \langle V(r)V(s)\rangle\langle y(s)\rangle ds, \quad \langle y(0)\rangle = y_o, \quad 0 \le r \le \tau_o/\varepsilon^2 \ .$$

Now we apply the <u>long-time-Markovian approximation</u> to (2.8) which means that we pull $\langle y(s)\rangle$ outside the integral in (2.8), evaluate it at $s = r$ and extend the integration to infinity. However, because of oscillatory factors as in (2.2) the integral to infinity will not exist and it must be replaced by

$$2.9 \qquad \overline{V} = \lim_{T\uparrow\infty} \frac{1}{T} \int_{r_o}^{r_o+T} \int_{r_o}^s \langle V(s)V(\sigma)\rangle d\sigma ds \ .$$

The long-time-Markovian approximation $\bar{y}(r)$ of $\langle y(r)\rangle$ is thus given by solving

$$2.10 \qquad \frac{d\bar{y}(r)}{dr} = \varepsilon^2 \bar{V}\,\bar{y}(r), \quad \bar{y}(0) = y_o \;.$$

We assume that the limit (2.9) exists and is independent of $r_o \geq 0$.

We shall employ here exclusively the approximation (2.10) because it yields results in their simplest and most useful form and because in the context considered here, the advantages that (2.8) may have over (2.10) are neutralized by its complexity.

Let us restate directly the connection between (2.3) and the approximation (2.10) as a formal asymptotic limit. Let

$$2.11 \qquad \tau = \varepsilon^2 r, \quad y^{\varepsilon}(\tau) = y(\tau/\varepsilon^2) \;.$$

Here τ is the scaled range relative to the size of the fluctuations and $y^{\varepsilon}(\tau)$ is the vector of mode amplitudes as functions of scaled ranges (with radiation neglected for simplicity in this section). We have that, as $\varepsilon \to 0$, $0 \leq \tau \leq \tau_o$, $\langle y^{\varepsilon}(\tau)\rangle$ tends to $\bar{y}(\tau)$ where

$$2.12 \qquad \frac{d\bar{y}(\tau)}{d\tau} = \bar{V}\,\bar{y}(\tau), \quad \bar{y}(0) = y_o \;,$$

with $\bar{V}$ defined by (2.9).

In the form (2.12) given here the above asymptotic limit can be given a rather complete mathematical treatment; see [10] and references to other work there as well as Stratonovich [11] and Khasminskii [12]. In fact, one can show that the error in the approximation $\bar{y}(\tau)$ is $O(\varepsilon)$, uniformly in $0 \leq \tau \leq \tau_o$ ($\tau_o < \infty$ but arbitrary). The condition on $V(r)$ that we mentioned was needed essentially to allow the transition from (2.6) to (2.7); it is called the mixing condition; we shall not give its technical meaning here ([10]). Physically, it means that the fluctuations $\mu(r,z)$ and $\mu(r+s,z')$ at two points separated in range by s tend to become statistically independent, in a sufficiently strong sense, as s becomes larger and larger. This is a perfectly acceptable assumption for the underwater sound propagation problem.

We show now how one can obtain the full statistical description of $y(r)$ in the asymptotic limit corresponding to (2.12). As we describe in [10], for example, the process $y^{(\varepsilon)}(\tau)$, defined by (2.11), converges as $\varepsilon \to 0$ to a Markov process $y^{(0)}(\tau)$, $0 \leq \tau \leq \tau_o$, so that it suffices to find the Fokker-Planck differential operator for the limiting process $y^{(0)}(\tau)$. Let us outline how the derivation of the Fokker-Planck equation follows the pattern (2.3) → (2.10) or (2.12).

Let $f(y)$ be a smooth function of N real variables $y = (y_1, y_2, \ldots, y_N)$. Let us solve (2.3) in the interval $[s,r]$, $0 \leq s \leq r$, with $y(s) = y_o$ given and let us denote the solution by $y(r,s;y_o)$. Define $\tilde{y}(r,s;y_o)$ by

$$\tilde{y}(r,s;y_o) = f(y(r,s;y_o)) . \tag{2.13}$$

By elementary computation we find that

$$\frac{\partial \tilde{y}(r,s;y_o)}{\partial s} + \varepsilon \sum_{p,q=1}^{N} V_{pq}(s) y_{oq} \frac{\partial}{\partial y_{op}} \tilde{y}(r,s;y_o) = 0 , \tag{2.14}$$

$$s < r , \qquad \tilde{y}(r,r;y_o) = f(y_o) .$$

Thus, (2.14) is formally again a problem of the same form as (2.3). Now however the independent variable is s and it runs backwards and the operator corresponding to V of (2.3) is a differential operator. Suffice it to say that this formal correspondence of objects can be carried all the way to obtain the asymptotics corresponding to (2.8), (2.10) or (2.12); see [6, 8, 11]

Let us now give the form of the Fokker-Planck operator corresponding to the limiting Markov process $y^{(0)}(\tau)$ and whose derivation follows the lines just sketched. Let $P(\tau,y;y_o)$ denote the transition probability density of $y^{(0)}(\tau)$ given $y^{(0)}(0) = y_o$, i.e.,

$$P(\tau,y;y_o)dy = P\{y^{(0)}(\tau) \in dy \mid y^{(0)}(0) = y_o\} .$$

Then, $P(\tau,y;y_o)$ satisfies the equation

$$2.15 \qquad \frac{\partial P(\tau,y;y_o)}{\partial \tau} = \sum_{p,q,p',q'=1}^{N} \frac{\partial^2}{\partial y_p \partial y_{p'}} (a_{pq,p'q'} y_q y_{q'} P) - \sum_{p,q=1}^{N} \frac{\partial}{\partial y_p} (b_{pq} y_q P) ,$$

$$\tau > 0 , \qquad P(0,y,y_o) = \delta(y-y_o) .$$

The diffusion coefficients $a_{pq,p'q'}$ and the drift coefficients b_{pq} are given by

$$2.16 \qquad a_{pq,p'q'} = \lim_{T\uparrow\infty} \frac{1}{T} \int_{t_o}^{t_o+T} \int_{t_o}^{s} \langle V_{pq}(\sigma) V_{p'q'}(s)\rangle d\sigma ds ,$$

$$b_{pq} = \lim_{T\uparrow\infty} \frac{1}{T} \int_{t_o}^{t_o+T} \int_{t_o}^{s} \sum_{q'=1}^{N} \langle V_{q'q}(\sigma) V_{pq'}(s)\rangle d\sigma ds .$$

These limits are assumed to exist independently of t_o, which is the case if $V_{pq}(r)$ is given by a formula such as (2.2) with $(\mu_{pq}(r))$ a stationary process.

In the following section we also employ the adjoint of the Fokker-Planck equation called the <u>backward Kolmogorov equation</u> for the Markov process $y^{(0)}(\tau)$. If $f(y)$ is a smooth function and if $y^{(0)}(0) = y_o$ then

$$2.17 \qquad u(\tau,y_o) = \langle f(y^{(0)}(\tau))\rangle = \int P(\tau,y;y_o) f(y) dy$$

satisfies the equation

$$\frac{\partial u(\tau,y_o)}{\partial \tau} = \sum_{p,q,p',q'=1}^{N} a_{pq,p'q'} y_{oq} y_{oq'} \frac{\partial^2 u(\tau,y_o)}{\partial y_{op}\partial y_{op'}} + \sum_{p,q=1}^{N} b_{pq} y_{oq} \frac{\partial u(\tau,y_o)}{\partial y_{op}} , \tag{2.18}$$

$$\tau > 0 , \qquad u(0,y_o) = f(y_o) .$$

The differential operator on the right hand side of (2.18) is called the <u>infinitesimal generator</u> of the Markov process $y^{(0)}(\tau)$.

We restate once again the approximation result we shall be using. Let $y(\tau)$ be the process defined by (2.3) and let $y^{(\varepsilon)}(\tau)$ be defined by (2.11). Then for any smooth function $f(y)$ we have that

$$\langle f(y^{(\varepsilon)}(\tau))\rangle \underset{\varepsilon\downarrow 0}{\longrightarrow} \langle f(y^{(0)}(\tau))\rangle = u(\tau,y_o), \qquad 0 \leq \tau \leq \tau_o , \tag{2.19}$$

where $u(\tau,y_o)$ satisfies (2.18) and the error in the approximation is $O(\varepsilon)$.

To obtain the asymptotic behavior of averages of $y^{(\varepsilon)}(\tau)$ at the different scaled ranges, τ_1 and τ_2 say, we use the Markov property of the limit process $y^{(0)}(\tau)$. The joint probability density of $y^{(0)}(\tau_1)$ and $y^{(0)}(\tau_2)$, $0 \leq \tau_1 \leq \tau_2$, is given by the product

$$P(\tau_2-\tau_1,y_2;y_1)P(\tau_1,y_1;y_o) .$$

so that if we know the solution $P(\tau,y;y_o)$ of (2.15) we can compute the approximations to averages of 2-range quantities.

Naturally solving (2.15) will turn out to be a very difficult problem. There is however a surprising amount of information one can obtain without solving the full equation. We should also remark that since the presently available mathematical theory referred to above is not sufficient for our problem (1.24), the above results will be applied formally in the following section.

3. Application of asymptotic methods to coupled mode equations

In this section we shall apply the asymptotic method described in the previous section to the system (1.24). The finite-dimensional vector $y(r)$ of Section 2 must now be replaced by the vector $c(r)$ of (1.22) which includes the continuously indexed radiation mode amplitudes. In addition, we must allow for the fact that $c(r)$ is complex-valued. For this purpose we consider jointly the vector $c(r)$ and its conjugate $c^*(r)$ which satisfies the complex conjugate of (1.24). Recall that the matrix $\mu(r)$ is real and symmetric. It is more convenient to deal with $c(r)$ and $c^*(r)$ rather than their real and imaginary parts.

Instead of writing the answer directly by applying, with appropriate modification, the formulas of the last section we shall proceed in a manner that exposes again the ideas in the derivation. The first step consists in obtaining here the analog of (2.14). Let $c(r,s;c_o)$ denote the solution of (1.24) with $s < r$ and initial condition $c(s,s;c_o) = c_o$, and let $c^*(r,s;c_o^*)$ denote its complex conjugate which satisfies the complex conjugate of (1.24). If $c = x + iy (i = \sqrt{-1})$ is a complex variable we define, as usual, complex derivatives as follows.

$$3.1 \qquad \frac{\partial}{\partial c} = \frac{1}{2}\left(\frac{\partial}{\partial x} - i\,\frac{\partial}{\partial y}\right), \qquad \frac{\partial}{\partial c^*} = \frac{1}{2}\left(\frac{\partial}{\partial x} + i\,\frac{\partial}{\partial y}\right) \quad ,$$

Let $f(c,c^*)$ be a smooth, real-valued function of its arguments and consider

$$3.2 \qquad u(r,s;c_o,c_o^*) = f(c(r,s;c_o),\ c^*(r,s;c_o^*)) \quad .$$

It is easily verified that u satisfies the following analog of (2.14).

$$3.3 \qquad \frac{\partial u(r,s;c_o,c_o^*)}{\partial s} + \varepsilon V(s) u(r,s;c_o,c_o^*) = 0 \ , \qquad s < r \ ,$$

$$u(r,r;c_o,c_o^*) = f(c_o,c_o^*) \quad .$$

Here $V(s)$ is the differential operator given by

3.4
$$V(s) = i \sum_{p,q=1}^{N} \mu_{pq} e^{i(\beta_q - \beta_p)s} c_{oq} \frac{\partial}{\partial c_{op}}$$

$$+ i \sum_{p=1}^{N} \int_0^{k^2} d\gamma \mu_{p\gamma}(s) e^{i(\sqrt{\gamma} - \beta_p)s} c_o(\gamma) \frac{\partial}{\partial c_{op}}$$

$$+ i \sum_{p=1}^{N} \int_0^{k^2} d\gamma \mu_{\gamma p}(s) e^{i(\beta_p - \sqrt{\gamma})s} c_{op} \frac{\delta}{\delta c_o(\gamma)}$$

$$+ i \int_0^{k^2} \int_0^{k^2} d\gamma d\gamma' \mu_{\gamma\gamma'}(s) e^{i(\sqrt{\gamma'} - \sqrt{\gamma})s} c_o(\gamma') \frac{\delta}{\delta c_o(\gamma)}$$

$$+ \text{ c.c.} \quad .$$

Here we have used the abbreviation c.c. to stand for complex conjugate of the preceding expression in the sum and we have denoted by $\delta/\delta c_o(\gamma)$ the variational derivative. The elementary formal calculus of variational derivatives will be employed without special comment in the sequel ([13]). In particular, complex variational derivatives are defined in the same way as (3.1).

We continue now with the asymptotic analysis of (3.3). According to the outline given in Section 2 we first rescale the problem using

3.5
$$\sigma = \varepsilon^2 s \,, \qquad \tau = \varepsilon^2 r \quad ,$$

to denote the scaled initial and final ranges respectively, and define the scaled complex mode amplitudes and u of (3.2) as a function of these scaled variables by

3.6
$$c^{(\varepsilon)}(\tau,\sigma;c_o) = c(\tau/\varepsilon^2,\sigma/\varepsilon^2;c_o)$$
$$u^{\varepsilon}(\tau,\sigma;c_o,c_o^*) = u(\tau/\varepsilon^2,\sigma/\varepsilon^2;c_o,c_o^*) \quad .$$

Then, as $\varepsilon \to 0$, $\langle u^{\varepsilon}(\tau,\sigma;c_o,c_o^*)\rangle$ tends to the function $\bar{u}(\tau-\sigma,c_o,c_o^*)$ which depends on $\tau-\sigma$ only and satisfies the backward Kolmogorov equation

$$3.7 \qquad \frac{\partial\bar{u}}{\partial\tau} = \bar{V}\,\bar{u}\ , \qquad \tau > 0\ , \qquad \bar{u}(0,c_o,c_o^*) = f(c_o,c_o^*)\quad .$$

The infinitesimal generator $\bar{V}$ is given by the formula

$$3.8 \qquad \bar{V} = \lim_{T\uparrow\infty} \frac{1}{T}\int_{r_o}^{r_o+T}\int_{r_o}^{s} \langle V(\sigma)V(s)\rangle d\sigma ds\quad ,$$

corresponding to (2.9) and with $V(s)$ given by (3.4).

In order to find the explicit form of the infinitesimal generator $\bar{V}$ of the limit Markov process $c^{(0)}(\tau;c_o,c_o^*)$, to which the process $c^{(\varepsilon)}(\tau;c_o,c_o^*)$ converges, we must insert (3.4) into (3.8) and perform the indicated ensemble averages and integration. This calculation is straightforward but lengthy so we shall omit it and write the result directly. To simplify the notation we shall drop the subscript o from c_o and c_o^*. We find that

$$3.9 \qquad \bar{V} = - \sum_{\beta_p-\beta_q=\beta_{q'}-\beta_{p'}} \left\{ \int_0^\infty \langle\mu_{pq}(t)\mu_{p'q'}(0)\rangle e^{i(\beta_q-\beta_p)t} dt \right\} c_q \frac{\partial}{\partial c_p}\left(c_{q'}\frac{\partial}{\partial c_{p'}}\right)$$

$$+ \sum_{\beta_p-\beta_q=\beta_{p'}-\beta_{q'}} \left\{ \int_0^\infty \langle\mu_{pq}(t)\mu_{p'q'}(0)\rangle e^{i(\beta_q-\beta_p)t} dt \right\} c_q c_{q'}^* \frac{\partial^2}{\partial c_p \partial c_{p'}^*}$$

$$- \sum_{\beta_p=\beta_q} \int_0^{k^2} d\gamma \left\{ \int_0^\infty \langle\mu_{\gamma q}(t)\mu_{p\gamma}(0)\rangle e^{i(\beta_q-\sqrt{\gamma})t} dt \right\} c_q \frac{\partial}{\partial c_p}$$

$$+ \sum_{\beta_p=\beta_q} \int_0^{k^2}\int_0^{k^2} d\gamma d\gamma' \left\{ \int_0^\infty \langle\mu_{\gamma q}(t)\mu_{\gamma p}(0)\rangle e^{-i(\beta_q-\sqrt{\gamma})t} dt \right\}$$

$$\cdot\ \Delta(\gamma-\gamma')c_q^* c_p \frac{\delta^2}{\delta c(\gamma')\delta c^*(\gamma)} + \text{c.c.}\quad .$$

Here the summations extend over all indices for which the indicated equalities hold, c.c. stands again for the complex conjugate of the first part of the operator $\overline{V}$ and $\Delta(\gamma)$ is defined as identically equal to zero for $\gamma \neq 0$ and equal to one at $\gamma = 0$. From this definition of $\Delta(\gamma)$ it follows that the term with the second functional derivative in (3.9) will be zero unless a delta function $\delta(\gamma-\gamma')$ appears from the differentiation.

We note that in the derivation of (3.9) we have employed the symmetry of the random matrix $\mu(r)$ with elements given in (1.19) and we have also employed the hypothesis stated below (1.2) that the random fluctuations of the refractive index $\mu(r,z)$ vanish for sufficiently large z.

The operator $\overline{V}$ of (3.9) has the following important property. If $f(c,c^*)$ is a function of the discrete components $c_1,c_2,\ldots,c_N,c_1^*,c_2^*,\ldots,c_N^*$ only, then $\overline{V}f(c,c^*)$ is also a function of $c_1,c_2,\ldots,c_n,c_1^*,c_2^*,\ldots,c_N^*$ only. This means that in the asymptotic limit under consideration the statistical properties of the propagating trapped mode amplitudes can be described independently of those of the radiation mode amplitudes. This decoupling of the propagating trapped modes from the radiation modes is a direct consequence of the assumption that the refractive index fluctuations $\mu(r,z)$ vanish for large z. The physical meaning of this decoupling is clear under this hypothesis because it is impossible for the inhomogeneities to cause energy transfer from the radiation modes into the trapped modes. Of course, energy can escape out of the trapped modes into the radiation modes and so get lost into the bottom of the ocean. This effect is due to the third term in the definition (3.9) of $\overline{V}$.

Because of the importance of the decoupling we restate the results of the asymptotic analysis again as follows. Let $c_T^{(\varepsilon)}(\tau) = (c_1^{(\varepsilon)}(\tau),\ldots,c_N^{(\varepsilon)}(\tau))$ be the complex-valued, random, <u>trapped-mode</u> amplitudes[†] at scaled range τ, as in (3.5), (3.6) and with $(c_1^{(\varepsilon)}(0),c_2^{(\varepsilon)}(0),\ldots,c_N^{(\varepsilon)}(0)) = (c_{o1},c_{o2},\ldots,c_{oN})$ given, nonrandom, initial mode amplitudes at range zero. Then, as $\varepsilon \to 0$ and the scaled range stays finite, the stochastic process $c_T^{(\varepsilon)}(\tau) = (c_1^{(\varepsilon)}(\tau),c_2^{(\varepsilon)}(\tau),\ldots,c_N^{(\varepsilon)}(\tau))$ converges to the diffusion Markov process $c_T(\tau) = (c_1(\tau),c_2(\tau),\ldots,c_N(\tau))$ with values in C^N (the complex

---

† The subscript T stands for "trapped".

N-dimensional space) whose Fokker-Planck operator $\overline{V}_T^*$ is the formal adjoint of $\overline{V}$ given by (3.9) with the fourth sum (the variational derivatives) omitted. Let $P(\tau,c_1,\ldots,c_N,c_1^*,\ldots,c_N^*;c_{o1},\ldots,c_{oN},c_{o1}^*,\ldots,c^*_{oN})$ denote the transition probability density of $(c_1(\tau),\ldots,c_N(\tau))$, i.e., the solution of the Fokker-Planck equation

$$3.10 \qquad \frac{\partial P}{\partial \tau} = \overline{V}_T^* P\,, \qquad P(0,c,c^*;c_o,c_o^*) = \delta(c-c_o)\delta(c^*-c_o^*)$$

with $\overline{V}_T^*$ defined as above and $c = (c_1,c_2,\ldots,c_N)$, etc. Then, as we mentioned at the end of Section 2, all statistical properties of $(c_1^\varepsilon(\tau),\ldots,c_N^\varepsilon(\tau))$ can be obtained in the limit $\varepsilon \to 0$ from the solution P of (3.10).

In order to study further the statistical properties of $c(\tau)$, the limiting mode amplitudes (with or without the radiation modes), we introduce the simplifying assumption of <u>nondegeneracy of the modes</u> as follows.

3.11 The propagation constants $\beta_1,\ldots,\beta_N$ are distinct along with their sums and differences.

Let us note that this assumption is violated when azimuthal fluctuations are present, i.e., $\mu = \mu(r,z,\theta)$. However, the results below can be recovered if we assume that the fluctuations are <u>statistically</u> rotationally invariant about the vertical axis at the source; see ([14]) for some comparable results. Thus, (3.11) is not as stringent as it may appear and we proceed to utilize it next.

In the nondegenerate case (3.11) the infinitesimal generator $\overline{V}$ of the limit Markov process $c(\tau)$ (with radiation modes included), given by (3.9), simplifies after some rearrangements to the following form

$$3.12 \qquad \overline{V} = \sum_{1\le q<p\le N} \left\{ \frac{1}{2} a_{pq}(A_{pq}A^*_{pq} + A^*_{pq}A_{pq}) + \tilde{a}_{pq}A_{pp}A^*_{qq} + i\hat{a}_{pq}(A_{qq}-A_{pp}) \right\}$$

$$+ \frac{1}{2}\sum_{p=1}^{N} \tilde{a}_{pp}A_{pp}A^*_{pp} - \sum_{p=1}^{N}\int_0^{k^2} d\gamma \left(\hat{\beta}_{p\gamma}c_p \frac{\partial}{\partial c_p} + \hat{\beta}^*_{p\gamma}c^*_p \frac{\partial}{c^*_p} \right)$$

$$+ \sum_{p=1}^{N} \int_0^{k^2} \int_0^{k^2} d\gamma d\gamma' \Delta(\gamma-\gamma') c_p c_p^* \left\{ \hat{b}_{p\gamma} \frac{\delta^2}{\delta c^*(\gamma')\delta c(\gamma)} + \hat{b}^*_{p\gamma} \frac{\delta^2}{\delta c(\gamma')\delta c^*(\gamma)} \right\}$$

Here

3.13 $$A_{pq} = c_p \frac{\partial}{\partial c_q} - c_q^* \frac{\partial}{\partial c_p^*} = - A^*_{qp}$$

and

3.14 $$a_{pq} = \int_{-\infty}^{\infty} \langle \mu_{pq}(t)\mu_{pq}(0)\rangle \cos(\beta_p-\beta_q)t \; dt$$

$$\hat{a}_{pq} = \int_0^{\infty} \langle \mu_{pq}(t)\mu_{pq}(0)\rangle \sin(\beta_p-\beta_q)t \; dt$$

3.15 $$\tilde{a}_{pq} = \int_{-\infty}^{\infty} \langle \mu_{pp}(t)\mu_{qq}(0)\rangle \; dt$$

3.16 $$\hat{b}_{p\gamma} = \int_0^{\infty} \langle \mu_{p\gamma}(t)\mu_{p\gamma}(c)\rangle e^{i(\beta_p-\sqrt{\gamma})t} \; dt \;, \qquad p,q=1,\ldots,N \;.$$

The infinitesimal generator for $c_T(\tau)$, the trapped mode amplitude limit process, is given by

3.17 $$\overline{V}_T = \sum_{1\le q<p\le N} \left\{ \frac{1}{2} a_{pq}(A_{pq}A^*_{pq} + A^*_{pq}A_{pq}) + \tilde{a}_{pq}A_{pp}A^*_{qq} + i\hat{a}_{pq}(A_{qq}-A_{pp}) \right\}$$

$$+ \frac{1}{2} \sum_{p=1}^{N} \tilde{a}_{pp}A_{pp}A^*_{pp} - \sum_{p=1}^{N} \left\{ \hat{b}_p c_p \frac{\partial}{\partial c_p} + \hat{b}^*_p c^*_p \frac{\partial}{\partial c^*_p} \right\}$$

where

$$3.18 \qquad \hat{b}_p = \int_0^{k^2} \hat{b}_{p\gamma} d\gamma .$$

Thus, the process $c_T(\tau)$, $\tau \geq 0$ is a diffusion Markov process with state space C^N , the complex N-dimensional space. Note that the coefficients (3.14), (3.15) are power spectra of the matrix elements of the fluctuation process $\mu(r,z)$.

In Sections 4 - 8 we study in detail properties of the process $c_T(\tau)$; second moments, fourth moments, its behavior as $N \uparrow \infty$, etc. All results herein follow from the form (3.17) of the infinitesimal generator. In Section 5 we discuss some simple generalizations to account for slow modulation effects not incorporated into (3.17). No essential changes are made there however.

In the remainder of this section we shall use the full operator (3.17) to show that when radiation is present the energy of the waves decreases to zero as the scaled range τ increases to infinity.

Assume that

$$3.19 \qquad 0 < \delta = \min_{1 \leq p \leq N} (\hat{b}_p + \hat{b}_p^*)$$

and let

$$|c|^2 = \sum_{p=1}^{N} c_p c_p^*$$

Then, it can be readily verified from (3.17) that

$$3.20 \qquad \overline{V}_T |c|^2 \leq -\delta |c|^2 .$$

The statement of the result is now this. For any $\varepsilon > 0$ and any starting value $c(0)$, say,

$$3.21 \qquad P\{|c_T(\tau)|^2 \leq (|c_T(0)|^2+\varepsilon)e^{-\delta\tau}, \text{ for all } \tau \geq 0\} \geq \frac{\varepsilon}{|c_T(0)|^2+\varepsilon} .$$

Thus, for $|c(0)|$ sufficiently small, the probability that $|c(\tau)|^2$ will decay exponentially fast as $\tau \to \infty$ can be made arbitrarily close to one.

Restated in more physical terms, we have that because of radiation losses the energy carried by the trapped modes will decay exponentially with range with probability as close to one as desired.

The demonstration of (3.21) requires some facts about stochastic differential equations and can be found in [15, p.325].

The validity of the forward scattering approximation, which we have adopted, can be assessed on the basis of the results of this section (and the previous one) applied to the full system of mode amplitudes c^+ and c^- . It is necessary that

$$3.22 \qquad \int_{-\infty}^{\infty} \langle\mu_{pq}(t)\mu_{pq}(0)\rangle \cos(\beta_p+\beta_q)t \, dt , \qquad p,q=1,\ldots,N ,$$

be negligibly small compared to a_{pq} of (3.14) along with similar relations for coupling to radiation. This is a useful condition for checking the validity of the forward scattering approximation in the stochastic context.

4. Coupled power equations

In Section 3 it was shown that the limiting Markov process $c(\tau)$ has a transition probability density satisfying (3.10). Once this density function p is known, a complete statistical description of the limiting behavior, as $\varepsilon \downarrow 0$, of the mode amplitudes $c^{\varepsilon}(\tau)$ is available.

In this section (and in section 6) we shall obtain information about $c_T(\tau)$[†] without actually solving for p. This is possible because $\overline{V}_T$ defined by (3.17) has the following special property. The coefficients of the second derivatives are

[†] Information about $c(\tau)$ not contained in $c_T(\tau)$, the trapped mode amplitudes, can be obtained by using the conservation relation (1.25).

homogeneous of degree 2 and the coefficients of the first derivatives are homogeneous of degree 1. Thus, we can obtain closed equations for moments of $c_T(\tau)$. Since we shall work exclusively with the trapped modes, we shall drop the subscript T from now on.

Let

4.1 $$W_r(\tau) = \int c_r c_r^* P(\tau,c,c^*;c_o,c_o^*)dcdc^*$$

$$= \lim_{\varepsilon\downarrow 0} <|c_r^{\varepsilon}(\tau)|^2> , \qquad \tau \geq 0 , \quad r=1,2,\ldots,N .$$

Using the equation

4.2 $$\frac{\partial P}{\partial \tau} = \overline{V} * P$$

and the form (3.17) of $\overline{V}$ we obtain after an elementary computation the <u>coupled power equations</u>:

4.3 $$\frac{dW_r(\tau)}{d\tau} = -b_r W_r(\tau) + \sum_{p=1}^{N} (a_{rp}W_p - a_{pr}W_r) , \quad \tau > 0$$

$$W_r(0) = |c_{or}|^2 , \quad r=1,2,\ldots,N .$$

Here a_{pq} is as in (3.14)

4.4 $$a_{pq} = \int_{-\infty}^{\infty} <\mu_{pq}(t)\mu_{pq}(0)> \cos(\beta_p-\beta_q)t \, dt , \qquad p,q=1,\ldots,N .$$

and b_p is given by

4.5 $$b_p = 2 \operatorname{Re} \left\{ \int_0^{k^2} \hat{b}_{p\gamma} d_\gamma \right\} = \int_0^{k^2} d\gamma \int_{-\infty}^{\infty} dt <\mu_{p\gamma}(\tau)\mu_{p\gamma}(0)> \cos(\beta_p - \sqrt{\gamma}) , \ p=1,\ldots,N.$$

We note that the energy transport coefficients a_{pq} are nonnegative, being power spectra, and symmetric and the radiation loss coefficients b_p are nonnegative, being integrals of power spectral functions.

If the smallest radiation loss coefficient is positive, then the solution of (4.3) tends to zero as $\tau \uparrow \infty$. This is elementary; in fact we have

$$\frac{d}{d\tau} \sum_{r=1}^{N} W_r(\tau) = - \sum_{r=1}^{N} b_r W_r(\tau) < - \min_p b_p \sum_{r=1}^{N} W_r(\tau)$$

from which an exponential decay is obtained.

On the other hand if $b_p = 0$, $p=1,2,\ldots,N$ then, $W_r(\tau)$ tends to equipartition as $\tau \uparrow \infty$:

$$\lim_{\tau\uparrow\infty} W_r(\tau) = \frac{1}{N} \sum_{p=1}^{N} |c_{op}|^2 , \quad r=1,2,\ldots,N.$$

Here we must use the symmetry of the coefficients a_{pq} which goes back to the symmetry of μ_{pq} .

To obtain the mode amplitude correlations at the different scaled ranges, we use the Markov property of the limit process $c(\tau)$. Thus, if

$$W_{rs}(\tau+\sigma,\tau) = \lim_{\varepsilon\downarrow 0} \langle c_r^{\varepsilon}(\tau+\sigma)c_s^{\varepsilon *}(\tau)\rangle \tag{4.6}$$

$$= \iint c_r \tilde{c}_s^* P(\sigma,c,c^*;\tilde{c},\tilde{c}^*) P(\tau,\tilde{c},\tilde{c}^*,c_o,c_o^*) dc\,dc^*\,d\tilde{c}\,d\tilde{c}^* ,$$

we find from (4.2) and (3.17) that (setting r=s for simplicity)

$$W_{rr}(\tau+\sigma,\tau) = W_r(\tau) \exp\left(- \sum_{p=1}^{N} \left[(\frac{1}{2} a_{rp} + i\hat{a}_{rp}) + \hat{b}_r\right] \sigma\right) \tag{4.7}$$

where $\hat{b}_p$ is defined by (3.16) and (3.18) and $a_{rp}, \hat{a}_{rp}$ are defined by (3.14). In (4.7) $W_r(\tau)$ are obtained by solving (4.3).

The coupled power equations (4.3) constitute a basic tool in studying energy transport in the ocean†. The reason is simply that equations (4.3) are

(i) relatively easy to solve and make sense intuitively,

(ii) depend on relatively few physical parameters,

(iii) yield information about quantities of direct physical interest: average mode powers

In implementing (4.3) one must decide what the coefficients a_{pq} and b_p are. If N is not too big ($N \leq 10$, say) i.e., at low frequency, the coefficients may be estimated from data. Although (4.3) are valid only asymptotically (cf.(4.1)), they are expected to give reasonable results under very general circumstances. Therefore the data used for the estimation need not be very deeply inside the theoretical region of validity of the asymptotics. Of course one may attempt to derive formulas for the coefficients a_{pq} and b_p by constructing theories for the fluctuation process μ in (1.1).

If N is large but the coefficients a_{pq} are negligibly small if $|p-q|>1$, then (4.3) can be approximated by a diffusion equation which again makes good physical sense and depends on an optimally small number of physical parameters. We consider this case in detail in section 8.

5. Quasi-static and slowly-varying coupled power equations

In this section we shall discuss the coupled power equations (4.3) in some detail. We shall introduce time dependence into (4.3) in a phenomenological manner and we shall study various limiting forms of the resulting equations.

It is clear that the coefficients a_{pq} and b_p defined by (4.4) and (4.5) need not be constants. They can be functions of the scaled range τ i.e., they can be slowly varying functions of the range. The coupled power equation (4.3) are valid as they stand with τ-dependent coefficients.

We note that the average mode powers $W_p(\tau)$ $p=1,\ldots,N$, are functions of the

† They have been used very effectively in fiber optics by Marcuse [16].

scaled range only. Let v_p be the group velocity of the pth mode (cf. below (1.11))

$$5.1 \qquad v_p = c_o \left[\frac{\partial \beta_p(k)}{\partial k} \right]^{-1} .$$

We shall assume that the average mode powers as functions of time and scaled range, $W_p(t,\tau)$, satisfy the equations

$$5.2 \qquad \frac{\partial W_p(t,\tau)}{\partial \tau} + \frac{1}{v_p} \frac{\partial W_p(t,\tau)}{\partial t} = \sum_{q=1}^{N} (a_{pq} W_q - a_{pq} W_p) - b_p W_p .$$

This is a reasonable extension of (4.3) and can be derived from first principles but we shall not do so here.

Equations (5.2) must be supplemented by initial and boundary conditions. Because of the nature of the approximations that led to (5.2) it is natural to suppose that $t \in (-\infty,\infty)$ and $\tau \geq 0$ so that

$$5.3 \qquad W_p(t,0) = W_{po}(t) , \qquad p=1,2,\ldots,N , \qquad -\infty < t < \infty ,$$

is given. This is the <u>time-pulse</u> problem. The corresponding <u>space-pulse</u> problem is defined for $t \geq 0$ and $-\infty < \tau < \infty$ with

$$5.4 \qquad W_p(0,\tau) = \tilde{W}_{po}(\tau) , \qquad p=1,\ldots,N , \qquad -\infty < \tau < \infty ,$$

given. Clearly (5.2) and (5.3) constitute the appropriate problem for us here.

The two problems (5.2), (5.3) and (5.2),(5.4) are dual to each other; both are well posed and the analysis below applies to both.

The physical meaning of the time-pulse problem is the following. The source (cf. (1.1)) is not precisely time harmonic with frequency ω but is a narrow band signal centered about ω . The input power of the source into the forward propagating modes is described by the function $W_{po}(t)$. In the approximation we are working here the average mode powers travel in space-time according to (5.2). In the absence of stochastic effects, $a_{pq} = 0$ and $b_p = 0$ (we assume no absorption), equation (5.2) tells us that the pulses launched at $\tau=0$ (range zero) propagate undistorted with the group velocity of the corresponding mode:

$$W_p(t,\tau) = W_{po}(\tau - v_p t) . \tag{5.5}$$

What is the effect of mode coupling upon the space-time behavior of the pulses $W_{po}(t)$, $p=1,\ldots,N$, $-\infty < t < \infty$, launched at range zero ($\tau=0$)? Even though (5.2), (5.3) is a realtively simple problem, well suited for numerical computations, it is not easy to get a general idea of what $W_p(t,\tau)$ looks like without additional assumptions. These assumptions fix the size of the terms in (5.2) relative to each other.

We shall examine two cases as follows.

(i) The terms $\frac{1}{v_p}\frac{\partial W_p}{\partial t}$ and $-b_p W_p$ are comparable to each other but the term

$$\sum_{q=1}^{N} (a_{pq}W_q - a_{pq}W_p) \tag{5.6}$$

is an order of magnitude larger.

(ii) The term $-b_p W_p$ is of order one, $\frac{1}{v_p}\frac{\partial W_p}{\partial t}$ is an order of magnitude bigger and the term (5.6) is an order of magnitude bigger than $\frac{1}{v_p}\frac{\partial W_p}{\partial t}$.

In both (i) and (ii) the coupling term (5.6) is assumed to play a predominant role. This is reasonable since, after all, mode coupling is what we want to analyze.

To describe (i) and (ii) we introduce a small parameter ε, not related to the parameter ε characterizing the size of the fluctuations in the index of refraction (cf. (1.1)).

Let us also denote by

$$s_p = \frac{-1}{v_p} \qquad p=1,2,\ldots,N, \tag{5.7}$$

the negative slowness of the modes. Assumption (i) corresponds to the following.

(i') a_{pq} is replaced by $\frac{1}{\varepsilon} a_{pq}$ but s_p and β_p remain $O(1)$ and we study (5.2), (5.3) in the limit $\varepsilon \to 0$, $\tau = O(1)$.

Similarly assumption (ii) corresponds to the following.

(ii') a_{pq} is replaced by $\frac{1}{\varepsilon^2} a_{pq}$, s_p is replaced by $\frac{1}{\varepsilon} s_p$ and b_p remains $O(1)$. We study (5.2), (5.3) in the limit $\varepsilon \to 0$, $\tau = O(1)$.

The analysis of (i') and (ii') is carried out by first and second order perturbation theory respectively. We follow the formulation of ([17] Theorems 1 and 2 p. 219) which fits precisely the needs of the present situation.

Define a_p by

5.8 $$a_p = \sum_{q \neq p} a_{pq} ,$$

and the matrix A by

5.9 $$A = \begin{bmatrix} -a_1 & a_{12} & \cdots & a_{1N} \\ a_{21} & -a_2 & \cdots & a_{2N} \\ \vdots & & & \\ a_{N1} & a_{N2} & \cdots & -a_N \end{bmatrix}$$

Define also

5.10 $$B = \begin{bmatrix} b_1 & 0 & \cdots & 0 \\ 0 & b_2 & & \\ \vdots & & \ddots & \\ 0 & & & b_N \end{bmatrix} ,$$

5.11 $$v = \begin{bmatrix} v_1 & 0 & \cdots & 0 \\ 0 & v_2 & & \\ \vdots & & \ddots & \\ 0 & & & v_N \end{bmatrix} , \quad S = \begin{bmatrix} s_1 & 0 & \cdots & 0 \\ 0 & s_2 & & \\ \vdots & & \ddots & \\ 0 & & & s_N \end{bmatrix} ,$$

5.12 $$P = \frac{1}{N}\begin{bmatrix} 1 & 1 & \cdots & 1 \\ 1 & 1 & & \\ \vdots & & \ddots & \\ 1 & & & 1 \end{bmatrix}$$

Let us first consider (i'). Then (5.2), (5.3) reduce to the following.

5.13 $$\frac{\partial W^{\varepsilon}}{\partial \tau} = \frac{1}{\varepsilon} AW^{\varepsilon} + S\,\frac{\partial W^{\varepsilon}}{\partial t} - BW^{\varepsilon} , \quad \tau > 0 ,$$

$$W^{\varepsilon}(t,0) = W_o(t) ,$$

where $W^{\varepsilon}(t,\tau)$ stands for the N-vector of the functions $W_p^{\varepsilon}(t,\tau)$. Now A has zero as eigenvalue which is isolated if we assume, as we do, that

5.14 $$a_{pq} > 0 , \quad p \neq q .$$

Moreover P is the projection matrix onto the one-dimensional eigenspace spanned by the eigenvector† $(1/N,\ldots,1/N)$ corresponding to the eigenvalue zero. It is easily seen (see the appendix) that if $\tau > 0$ i.e., we are away from the source region (as we must be anyway for reasons explained in Section 1) then

5.15 $$\lim_{\varepsilon \downarrow 0} W_p^{\varepsilon}(t,\tau) = \overline{W}(t,\tau) , \quad \tau > 0 ,$$

where the function $\overline{W}(t,\tau)$ satisfies

5.16 $$\frac{\partial \overline{W}}{\partial \tau} = + \overline{S}\,\frac{\partial \overline{W}}{\partial t} - \overline{b}\,\overline{W} , \quad \tau > 0$$

$$\overline{W}(t,0) = \frac{1}{N}\sum_{p=1}^{N} W_{op}(t) \equiv \overline{W}_o(t) , \quad -\infty < t < \infty ,$$

† $a_{pq} = a_{qp}$.

and

5.17 $$\bar{S} = \frac{1}{N} \sum_{p=1}^{N} s_p \qquad \text{(cf. (5.7))}$$

5.18 $$\bar{b} = \frac{1}{N} \sum_{p=1}^{N} b_p$$

If we define

5.19 $$\bar{v} = -\frac{1}{\bar{S}} ,$$

then the solution of (5.16) is

5.20 $$\bar{W}(t,\tau) = e^{-\bar{b}\tau}\, \bar{W}_o(\tau - \bar{v}t) .$$

The conclusions are as follows. In case (i') and away from a neighborhood of the source, in the limit of strong mode coupling, the pulse shape of each mode power function tends to the same function (5.20) which displays damping with distance from the source (coupling to radiation) and progagation with speed $\bar{v}$ which is given by

5.21 $$\bar{v} = \left[\frac{1}{N} \sum_{p=1}^{N} \frac{1}{v_p} \right]^{-1} \qquad \text{(cf. (5.7), (5.17), (5.19))}$$

i.e., the harmonic mean of the group velocities.

Let us also consider case (ii'). Equation (5.2), (5.3) becomes

5.22 $$\frac{\partial W^{\varepsilon}}{\partial \tau} = \frac{1}{\varepsilon^2} A W^{\varepsilon} + \frac{1}{\varepsilon} S \frac{\partial W^{\varepsilon}}{\partial t} - B W^{\varepsilon} , \qquad \tau > 0 ,$$

$$W^{\varepsilon}(t,0) = W_o(t) , \qquad -\infty < t < \infty ,$$

which is analogous to (5.13) Again we shall leave details to the appendix and discuss the results of the asymptotic analysis.

Let $\sigma^2 > 0$ be defined by (I = identity matrix)

$$5.23 \qquad \frac{1}{2}\sigma^2 I = - P(S - PS)A^{-1} (S - PS)P .$$

Note that A^{-1} is well defined, despite the fact that A has zero as an eigenvalue, because it acts on elements that have no components in the nullspace of A. Unfortunately one can not be more explicit about the determination of σ^2 since, in general, A^{-1} is not given explicitly. The computation is elementary, however. Let $\bar{b}$ and $\bar{v}$ be as before.

The result is that for $\tau > 0$ and fixed, $W_p^\varepsilon(t,\tau)$ behaves as $\varepsilon \to 0$ like the solution of

$$5.24 \qquad \frac{\partial \overline{W}^\varepsilon(t,\tau)}{\partial \tau} + \frac{1}{\varepsilon \bar{v}} \frac{\partial \overline{W}^\varepsilon(t,\tau)}{\partial t} = \frac{\sigma^2}{2} \frac{\partial^2 \overline{W}^\varepsilon(t,\tau)}{\partial t^2} - \bar{b}\, \overline{W}^\varepsilon(t,\tau) , \qquad \tau > 0 ,$$

$$\overline{W}^{\,\varepsilon}(t,0) = \overline{W}_o(t) = \frac{1}{N} \sum_{p=1}^{N} W_{op}(t) , \qquad -\infty < t < \infty .$$

More specifically, we have that $|W_p^\varepsilon - \overline{W}^{\,\varepsilon}| \to 0$ as $\varepsilon \to 0$ for all $p=1,2,\ldots,N$, $\tau > 0$ fixed and uniformally in $-\infty < t < \infty$. Note that, as in (5.13), (5.16), the asymptotic pulse form is the same for all modes but now we have diffusive pulse spreading in time due to the term $\frac{1}{2}\sigma^2\, \partial^2/\partial t^2$.

To get a better feeling for the nature of the approximation (5.24) let us assume that

$$5.25 \qquad \overline{W}_o(t) = a \frac{e^{-t^2/2\gamma^2}}{\sqrt{2\pi\gamma^2}} ,$$

i.e., a Gaussian pulse of width γ . Then,

$$5.26 \qquad \overline{W}^{\,\varepsilon}(t,\tau) = a\,\frac{e^{-(t-\frac{1}{\varepsilon\bar{v}}\tau)^2/2(\gamma^2+\tau\sigma^2)}}{\sqrt{2\pi(\gamma^2+\tau\sigma^2)}}\,e^{-\bar{b}\tau}$$

from which we conclude that the pulse spreading factor is

$$5.27 \qquad \left[1+\tau\frac{\sigma^2}{\gamma^2}\right]^{1/2} \quad .$$

The interesting thing about this conclusion is that pulse spreading of the time-pulse is proportional to the <u>square root</u> of the distance from the source.

Let us elaborate on this further†. Given that many modes propagate and they couple strongly, pulse spreading is proportional to the square root of the distance from the source and not proportional to τ as would be the case in the absence of mode coupling. Thus random mode coupling has a certain beneficial effect on signal propagation. Of course, it is assumed that all approximations that led to (5.26) are satisfied reasonably well (in particular (ii') above).

Appendix. First and second order perturbation theory for Boltzmann-like equations

Instead of giving a general treatment, like in [17], we shall deal with finite dimensional matrices which avoid technicalities but display all the features of the problem. In particular, we shall give a coordinate free analysis, independent of spectral theory so the results make sense in great generality††.

Let B be an N×N matrix such that

$$5A.1 \qquad e^{Bt} \longrightarrow P , \qquad t \uparrow \infty ,$$

† This is the effect discovered by Personik [18] and explained as above by Marcuse [19].

†† The notation that follows differs from the one of section 5 but agrees with [17] (where the terminology "Boltzmann-like" is explained).

where P is a projection matrix, projecting into the null space of B . Let A and C be N×N matrices.

Theorem 1 (First order perturbation theory)

For any $t > 0$ and any N-vector f

5A.2 $$e^{(B/\varepsilon+A)t} f \to e^{PAPt} Pf ,$$

provided (with C a constant)

5A.3 $$|e^{(B/\varepsilon+A)t} f| \leq C|f| .$$

Proof.

From the identity

5A.4 $$e^{(B/\varepsilon+A)t} = e^{Bt/\varepsilon} + \int_0^t e^{(B/\varepsilon+A)(t-s)} Ae^{Bs/\varepsilon} ds$$

and (5A.1) it follows that we must show

5A.5 $$e^{(B/\varepsilon+A)t} Pf \to e^{PAPt} Pf , \qquad 0 \leq t \leq T < \infty ,$$

Let

$$u^{\varepsilon}(t) = e^{(B/\varepsilon+A)t} Pf ,$$

$$\bar{u}(t) = e^{PAPt} Pf ,$$

and

5A.6 $$\bar{v}(t) = - B^{-1}(A-PA)\bar{u}(t) .$$

Note that B^{-1} is well defined since $P(A-PA) = 0$ (Fredholm alternative).

We have that

$$\left(\frac{B}{\varepsilon} + A - \frac{\partial}{\partial t}\right)\left(u^{\varepsilon}(t)-\bar{u}(t)-\varepsilon\bar{v}(t)\right)$$

$$= -\left(\frac{B}{\varepsilon} + A - \frac{\partial}{\partial t}\right)\left(\bar{u}(t) + \varepsilon\bar{v}(t)\right)$$

$$= - \left\{ B\bar{v}(t) + \left(A - \frac{\partial}{\partial t}\right)\bar{u}(t) + \varepsilon\ AB^{-1}(A-PA)\bar{u}(t) - B^{-1}(A-PA)PAP\bar{u}(t) \right\}$$

$$= O(\varepsilon)$$

Hence indeed ($|\cdot|$ is Euclidean norm) (with Pf=f)

5A.7 $$\sup_{0\le t\le T<\infty} |u^{\varepsilon}(t)-\bar{u}(t)| = O(\varepsilon)\ .$$

For this last conclusion we make use of (5A.3) clearly. The proof is complete.

Note that if B has also pure imaginary eigenvalues (5A.2) still holds provided Pf=f i.e., the starting vector is in the null space. The "initial layer" behavior, however, requires no oscillatory mode.

<u>Theorem 2</u> (Second order perturbation theory).

For any t > 0 and any N-vector f

5A.8 $$\lim_{\varepsilon\downarrow 0} \left| e^{(B/\varepsilon^2+A/\varepsilon+C)t}\ f - e^{(PAP/\varepsilon+V+PCP)t}Pf \right| = 0\ ,$$

where

5A.9 $$V = - P(A-PA)B^{-1}(A-PA)P = - PAB^{-1}(A-PA)P$$

and it is assumed that ($\tilde{C}$ a constant)

5A.10 $$\left| e^{(B/\varepsilon^2+A/\varepsilon+C)t}f \right| \le \tilde{C}|f|.$$

and

$$\left| e^{(PAP/\varepsilon+V+PCP)t} f \right| \leq C|f|. \tag{5A.11}$$

Remark 1. Note that the matrix $\frac{PAP}{\varepsilon} + V + PCP$ has the dimensions of the null space of B and hence it is much smaller than N×N, in general. Assumption (5A.10) is an a-priori estimate which in many cases of interest is easily obtained.

Remark 2. The terminology is not standard but we are, in fact, concerned with second order perturbation theory. The result (5A.8) is a nice way to express the answers in a general coordinate free way. For the case PAP $\equiv$ 0 see [20].

Remark 3. If PAP has only pure imaginary eigenvalues, then one can show easily that

$$\left| e^{-PAPt/\varepsilon} e^{(PAP/\varepsilon+V+PCP)t} Pf - e^{(\overline{V}+\overline{PCP})t} \right| = O(\varepsilon) \tag{5A.12}$$

where

$$\overline{V} + \overline{PCP} = \lim_{T\uparrow\infty} \frac{1}{T} \int_0^T e^{-PAPs} (V+PCP) e^{PAPs}\, ds. \tag{5A.13}$$

This is nothing but the method of averaging and the proof of (5A.12) is elementary.

Remark 4.

If f is replaced by Pf in (5A.8) then, it will be seen below, the error is O(ε).

Proof of Theorem 2.

By a preliminary argument as in Theorem 1 we can show that it is enough to consider $f = Pf$ if $t > 0$ i.e., we may drop the initial layers.

Let

$$u^{\varepsilon}(t) = e^{(B/\varepsilon^2+A/\varepsilon+C)t}\, Pf$$

$$\bar{u}^{\varepsilon}(t) = e^{(PAP/\varepsilon+V+PCP)t}\, Pf$$

We look for an expansion of the form

$$u^{\varepsilon} = \bar{u}^{\varepsilon} + \varepsilon \bar{v}_1^{\varepsilon} + \varepsilon^2 \bar{v}_2^{2} + \dots$$

with $\bar{v}_1^{\varepsilon}(t)$ and $\bar{v}_2^{\varepsilon}(t)$ to be determined so that

5A.15 $$\left(\frac{B}{\varepsilon^2} + \frac{A}{\varepsilon} + C - \frac{\partial}{\partial t}\right)(u^{\varepsilon} - \bar{u}^{\varepsilon} - \varepsilon \bar{v}_1^{2} - \varepsilon^2 \bar{v}_2^{\varepsilon}) = O(\varepsilon)$$

From this and (5A.10) the result follows provided $\bar{v}_1^{\varepsilon}$ and $\bar{v}_2^{\varepsilon}$ are bounded independently of ε. For this we use (5A.11) as will be seen next.

The left side of (5A.15) equals

$$-\left(\frac{B}{\varepsilon^2} + \frac{A}{\varepsilon} + C - \frac{\partial}{\partial t}\right)(\bar{u}^{\varepsilon} + \varepsilon \bar{v}_1^{\varepsilon} + \varepsilon^2 \bar{v}_2^{\varepsilon}) =$$

$$-\Bigg[\frac{1}{\varepsilon}(B\bar{v}_1^{\varepsilon} + A\bar{u}^{\varepsilon} - PAP\bar{u}^{\varepsilon})$$

$$+ (B\bar{v}_2^{\varepsilon} + A\bar{v}_1^{\varepsilon} + C\bar{u}^{\varepsilon} - (V+PCP)\bar{u}^{\varepsilon} - \varepsilon \frac{\partial \bar{v}_1^{\varepsilon}}{\partial t}\Bigg] + O(\varepsilon)$$

Now we choose

$$\bar{v}_1^{\varepsilon}(t) = -B^{-1}(A-PA)\bar{u}^{\varepsilon}(t) .$$

With this choice and (5A.9) it follows that $\bar{v}_2^{\varepsilon}(t)$ may be chosen (Fredholm alternative) as

$$\bar{v}_2^{\varepsilon}(t) = -B^{-1}\Bigg[\Big(-AB^{-1}(A-PA) + C - V - PCP\Big)\bar{u}^{\varepsilon}(t)$$

$$+ B^{-1}(A-PA)PAP\bar{u}^{\varepsilon}(t)\Bigg] .$$

With these choices (5A.15) follows provided $\bar{u}^{\varepsilon}(t)$ is bounded independently by ε as (5A.11) implies. The proof is complete.

Let us finally remark that the above demonstration is given in a "mathematicians" form i.e. in the opposite order in which one first obtains such a result. The steps can be turned around easily, however. The reason that we have given these elementary results (in Linear Algebra, essentially) in detail here is that they have broader significance and, in fact, they are easily modified to produce all the results of section 2 as is done, for example, in [21] .

6. Coupled fluctuation equations

The coupled power equations (4.3) derived in section 4 describe the solution for the average mode powers in the limit of section 2. For a given realization of the random fluctuation field $\mu(r,z)$, however, the modal powers may exhibit behavior substantially different from that of their statistical averages. Therefore, it is important to have some quantitative measure characterizing how far a given realization of power content in a mode deviates from its statistical average. Obviously, if one can solve (3.10) explicitly, one would have a complete probabilistic description of the modes as random functions of range (in the usual asymptotic limit). However, (3.10) is too difficult to solve explicitly and we only want information about power fluctuation. So we settle for the quantity

$$6.1 \qquad < [c_r^{(\varepsilon)}(\tau) c_r^{(\varepsilon)*}(\tau)]^2 > -[< c_r^{(\varepsilon)}(\tau) c_r^{(\varepsilon)*}(\tau)>]^2 \quad ,$$

$$r=1,2,\ldots,N \ ,$$

in the limit $\varepsilon \to 0$ with τ fixed. In this section we shall derive equations for second moments of modal powers. Using these coupled fluctuation equations and (4.3) we can then calculate (6.1).

Define $U_{rs}(\tau)$ by

$$6.2 \qquad U_{rs}(\tau) = \lim_{\varepsilon \downarrow 0} \langle |c_r^{(\varepsilon)}(\tau)|^2 \; |c_s^{(\varepsilon)}(\tau)|^2 \rangle$$

$$= \int c_r c_r^* c_s c_s^* \; P\,(\tau, c, c^*; c_o, c_o^*) dc dc^*$$

Using (3.10) and (3.12) we derive, as in section 4, the following equations for $U_{rs}(\tau)$.

$$6.3 \qquad \frac{dU_{rr}(\tau)}{d\tau} = - 2b_r U_{rr}(\tau) + 2 \sum_{\substack{p=1 \\ p \neq r}}^{N} a_{rp}[2U_{rp} - U_{rr}] \, ,$$

$$r = 1, 2, \ldots, N \, ,$$

$$\frac{dU_{rs}(\tau)}{d\tau} = - [b_r + b_s + 2a_{rs}]U_{rs} + \sum_{p=1}^{N} a_{rp}(U_{ps} - U_{rs})$$

$$+ \sum_{p=1}^{N} a_{ps}(U_{rp} - U_{rs}) \, ,$$

$$1 \leq r,s \leq N \, , \quad r \neq s$$

$$U_{rs}(0) = |c_{or}|^2 \; |c_{os}|^2 \, , \qquad 1 \leq r,s \leq N \, .$$

Here a_{pq} and b_p are defined by (4.4) and (4.5).

Clearly the quantity of interest (6.1) is given by

$$\left[U_{rr}(\tau) - W_r^2(\tau) \right]^{1/2} \, , \qquad 1 \leq r \leq N \, ,$$

where $W_r(\tau)$ satisfies (4.3). Thus, it suffices to solve (6.3).

To gain some insight into the structure of the equations (6.3) we shall consider some simple cases.

For a single guided (trapped) mode (N=1) it is clear that

$$6.4 \qquad U_{11}(\tau) = e^{-2b_1\tau}\, U_{11}(0) = W_1^2(\tau)$$

where $U_{11}(0) = |c_{o1}|^4$. In this case, the expected value of the power in the single mode decays exponentially with range as the energy is scattered by the random inhomogeneities and lost as radiation into the ocean floor. Note that the fluctuation $(U_{11}(\tau) - W_1^2(\tau))^{1/2} = 0$ i.e., the power decays to zero with probability one (cf. (3.21)).

Let us also consider the case of two guided modes (N=2). For simplicity we shall assume that the radiation loss coefficients b_1 and b_2 are equal; $b_1 = b_2 = b$. Also, since $a_{12} = a_{21}$ is the only coefficient that appears we shall call it a . Then, (6.3) becomes

$$6.5 \qquad \frac{d}{d\tau}\begin{bmatrix} U_{11} \\ U_{12} \\ U_{22} \end{bmatrix} = -\,2b \begin{bmatrix} 1 & 0 & 0 \\ 0 & 1 & 0 \\ 0 & 0 & 1 \end{bmatrix}\begin{bmatrix} U_{11} \\ U_{12} \\ U_{22} \end{bmatrix} + a \begin{bmatrix} -2 & 4 & 0 \\ 1 & -4 & 1 \\ 0 & 4 & -2 \end{bmatrix}\begin{bmatrix} U_{11} \\ U_{12} \\ U_{22} \end{bmatrix}$$

The solution of (6.5) is:

$$6.6 \qquad \begin{bmatrix} U_{11}(\tau) \\ U_{12}(\tau) \\ U_{22}(\tau) \end{bmatrix} = \frac{e^{-2b\tau}}{6}\begin{bmatrix} [2+3e^{-2a\tau}+e^{-6a\tau}] & [4-4e^{-6a\tau}] & [2-3e^{-2a\tau}+e^{-6a\tau}] \\ [1-e^{-6a\tau}] & [2+4e^{-6a\tau}] & [1-e^{6a\tau}] \\ [2-3e^{-2a\tau}+e^{-6a\tau}] & [4-4e^{-6a\tau}] & [2+3e^{-2a\tau}+e^{-6a\tau}] \end{bmatrix}\begin{bmatrix} U_{11}(0) \\ U_{12}(0) \\ U_{22}(0) \end{bmatrix}$$

Note that $U_{ij}(\tau) \xrightarrow{\tau\uparrow\infty} 0$, as should be by (3.21). In the absence of radiation loss i.e., b=0 , we obtain

$$6.7 \qquad \lim_{\tau\uparrow\infty} U_{11}(\tau) = \lim_{\tau\uparrow\infty} U_{22}(\tau) = 2 \lim_{\tau\uparrow\infty} U_{12}(\tau) = \tfrac{1}{3}(|c_{o1}|^2 + |c_{o2}|^2)^2 .$$

Since $W_r(\tau) \to \frac{1}{N} \sum_{p=1}^{N} |c_{op}|^2$ as $\tau \to \infty$ it follows that

$$6.8 \qquad \lim_{\tau\uparrow\infty} (U_{11}(\tau) - W_1^2(\tau))^{1/2} = \lim_{\tau\uparrow\infty} (U_{22}(\tau) - W_2^2(\tau))^{1/2} = \frac{|c_{o1}|^2 + |c_{o2}|^2}{\sqrt{12}} .$$

Let us next consider the general case of N trapped modes in the absence of radiation losses. We observe that setting

$$6.9 \qquad U_{rr} = 2U_{rs} = \text{constant} \qquad 1 \leq r,s \leq N , \; r \neq s$$

determines a critical point for (6.3) (with $b_1 = b_2 = \ldots = b_N = 0$). This point is asymptotically stable if $a_{pq} > 0$, $p \neq q$, and this implies that the constant is given by

$$6.10 \qquad \frac{2}{N(N+1)} \sum_{r=1}^{N} |c_{or}|^2$$

which generalizes (6.7).

From (6.10) and the equipartition result for $W_r(\tau)$ it follows that

$$6.11 \qquad \lim_{\tau\uparrow\infty} \frac{U_{rr}(\tau)}{W_r^2(\tau)} = 2 \lim_{\tau\uparrow\infty} \frac{U_{rs}(\tau)}{W_r(\tau)W_s(\tau)} = 2 \frac{N}{N+1} ,$$

$$1 \leq r,s \leq N , \quad r \neq s .$$

Consequently, the normalized covariance of the modal powers approaches the following limit as $\tau \uparrow \infty$

$$6.12 \qquad \lim_{\tau\uparrow\infty} \frac{U_{rs}(\tau) - W_r(\tau)W_s(\tau)}{W_r(\tau)W_s(\tau)} = \frac{-1}{N+1} , \qquad 1 \leq r,s \leq N , \qquad r \neq s .$$

From this it follows that for N large the cross correlation among mode powers becomes relatively less important.

Another consequence of (6.11) is the relation

$$6.13 \qquad \lim_{\tau\uparrow\infty} \frac{U_{rr}(\tau)-W_r^2(\tau)}{W_r^2(\tau)} = \frac{N-1}{N+1} \to 1 \quad \text{as} \quad N \uparrow \infty ,$$

which says that the <u>relative</u> fluctuations become large (=1) as $N \to \infty$.

We close this section by explaining why we feel it is important to study the dynamics of mode power exchange in the absence of radiation even though radiation losses are indeed a fact of life. It will become apparent in section 8, where a high frequency limit is taken $(N \to \infty)$, that if the random fluctuations in the transverse or depth direction are not too severe, radiation loss is negligible for many of the lower order trapped (or guided) modes. Only the higher order modes whose transverse wavenumbers are close to the edge of the propagating wavenumber band will be able to couple energy into the radiating modes. Thus, energy initially imparted to the lower order modes must diffuse through the guided (or trapped) modes and migrate to the band edge before it can couple into the radiation spectrum. That is why it is important to understand the nonradiating case.

7. Depth dependent quantities

In this section we point out, very briefly, that information about mode power statistics can be used directly to obtain information about the pressure field.

Consistent with the various approximations of section 1, from (1.12) and (1.16) we have

$$7.1 \qquad p(\tau/\varepsilon^2,z) = \sum_{r=1}^{N} \frac{e^{i\beta_r\tau/\varepsilon^2}}{\sqrt{\beta_r}} c_r(\tau/\varepsilon^2)\nu_r(z)$$

$$+ \int_0^{k^2} \gamma^{-1/4} e^{i\sqrt{\gamma}\,\tau/\varepsilon^2} c(\gamma,\tau/\varepsilon^2)\nu(z,\gamma)d\gamma$$

We shall concentrate on that part of the pressure that is due to the trapped modes and set

$$p^{(\varepsilon)}(\tau,z) = \sum_{r=1}^{N} \frac{e^{i\beta_r\tau/\varepsilon^2}}{\sqrt{\beta_r}} c_r^{(\varepsilon)}(\tau)\nu_r(z) \quad . \tag{7.2}$$

Clearly

$$|p^{(\varepsilon)}(\tau,z)|^2 = \sum_{r,s=1}^{N} \frac{e^{i(\beta_r-\beta_s)\tau/\varepsilon^2}}{\sqrt{\beta_r\beta_s}} c_r^{(\varepsilon)}(\tau)c_s^{(\varepsilon)*}(\tau)\nu_r(z)\nu_s(z) \ . \tag{7.3}$$

If we choose $\Delta > 0$ so that Δ is small compared to changes in $W_r(\tau)$ of (4.3) but Δ/ε^2 is large while $\beta_r - \beta_s$ is strictly positive for $r \neq s$, it follows that

$$\lim_{\varepsilon\downarrow 0} \frac{1}{\Delta} \int_{\tau}^{\tau+\Delta} \langle|p^{(\varepsilon)}(\sigma,z)|^2\rangle d\sigma = \sum_{r=1}^{N} \frac{1}{\beta_r} W_r(\tau)\nu_r^2(z) \ . \tag{7.4}$$

The meaning of (7.4) is this. The ensemble average of the squared modulus of the pressure as a function of depth and range, smoothed in range by a moving average, behaves, in the limit of weak fluctuations and long ranges, like the right hand side of (7.4) where W_r is determined by (4.3) and $\nu_r(t)$ are modes (cf. (1.7) - (1.10)).

8. High frequency approximation to the coupled power equations

In ([22]), Chernov uses a geometrical optics formulation to develop a ray diffusion equation for propagation in a random medium. This equation, satisfied by the probability density function, is essentially the heat equation, where time and space are replaced by range and ray angle, respectively. This equation is very appealing to the intuition since we expect that a highly directional acoustic excitation would spread out or diffuse in angle as range increases due to scattering by the random inhomogeneities. This ray diffusion model has been applied to acoustic propagation in randomly inhomogeneous oceans by Weinberg and Mellin [23] , [24].

The purpose of this section is to illustrate how one can arrive at a diffusion model by applying an appropriate limit to the stochastic framework and coupled power equations that we have derived. We are motivated and guided by the work of Gloge [25] and Marcuse [17] , [26] in the area of electromagnetic propagation in optical fibers. For definiteness, we shall consider a representative model problem and determine the behavior of the coupled power equations (4.6) in the limit where frequency (and thus the number of guided modes, N) increases.

The model problem that we shall consider is the slab configuration shown in Figure 3. The average soundspeed is assumed to have the piecewise-constant form shown and we assume that the random soundspeed fluctuations are confined to the ocean region $0 \le z \le d$, i.e. $\mu(r,z) = 0$ if $z > d$. For simplicity, we shall assume that the density of the medium is constant for $0 \le z < \infty$. Then, noting (1.4) and (1.5), we demand that the scaled acoustic pressure $p(r,z)$ satisfy:

8.1 $$\frac{\partial^2}{\partial r^2} p + \left[\frac{\partial^2}{\partial z^2} + k^2 n^2(z) \right] p + \varepsilon k^2 \mu(r,z) p = 0$$

$$p(r,o) = 0$$

$$p(r,z) \text{ and } \frac{\partial}{\partial z} p(r,z) \text{ continuous across } z = d$$

where $k = \omega/c_o = 2\pi f/c_o$ and the mean index of refraction $n(z)$ is:

8.2 $$n(z) \equiv \begin{cases} n_1 > 1 , & 0 \le z \le d \\ 1 , & z > d \end{cases}$$

This problem is similar to that modelling the propagation of electromagnetic waves in a dielectric slab whose refractive index is randomly perturbed. In fact, the texts of Marcuse, [16] , [26] , provide an excellent reference for the present discussion.

We begin by deriving the mode functions. For the bound or guided modes, we must solve the eigenvalue problem (c.f. (1.7)):

8.3 $$\frac{d^2}{dz^2}\nu_p(z) + k^2n^2(z)\nu_p(z) = \beta_p^2\nu_p(z)\,, \qquad p=1,\ldots,N$$

$$\nu_p(0) = 0$$

$$\nu_p \text{ and } \frac{d\nu_p}{dz} \text{ continuous across } z = d\,.$$

$$\int_0^\infty |\nu_p(z)|^2 dz = 1$$

where n(z) is defined by (8.2). Define:

8.4 $$\kappa_p \equiv \sqrt{\beta_p^2 - k^2} \qquad , \qquad \sigma_p = \sqrt{n_1^2k^2 - \beta_p^2}$$

The eigenvalue equation for β_p then becomes:

8.5 $$\sigma_p \cos\sigma_p d + \kappa_p \sin\sigma_p d = 0 \quad \text{or} \quad \tan\sigma_p d = \frac{-\sigma_p}{\kappa_p} \quad , \quad k < \beta_p < n_1 k$$

Notice that the number of guided modes, N, is determined by the number of solutions to (8.5) that exist within the range $k < \beta_p < n_1k$. This integer N is, therefore, an increasing function of frequency or wavenumber. The eigenfunctions $\nu_p(z)$ have the form:

8.6 $$\nu_p(z) = \begin{cases} A_p \sin\sigma_p z\,, & 0 \le z \le d \\ A_p \sin\sigma_p d\, e^{-\kappa_p(z-d)}\,, & d \le z < \infty \end{cases}$$

$$A_p \equiv \sqrt{2}\left[\frac{\sin^2\sigma_p d}{\kappa_p} + d\left(1 - \frac{\sin 2\sigma_p d}{2\sigma_p d}\right)\right]^{-1/2}$$

For the radiation modes, i.e. the mode functions corresponding to this continuous spectrum, we solve the transverse problem (c.f. (1.8)):

$$\frac{d^2}{dz^2}\nu(\gamma,z) + k^2 n^2(z)\nu(\gamma,z) = \gamma\nu(\gamma,z) \quad , \quad 0 < \gamma \leq k^2$$

8.7 $$\nu(\gamma,0) = 0$$

$$\nu(\gamma,z) \text{ and } \frac{d\nu}{dz}(\gamma,z) \text{ continuous across } z = d .$$

$$\int_0^\infty \nu(\gamma,z)\nu^*(\lambda,z)dz = \delta(\gamma-\lambda)$$

where $\delta(\cdot)$ denotes the Dirac delta function. There is no eigenvalue relation associated with this problem. A continuum of solutions exist, one for each value of γ in the range $0 < \gamma \leq k^2$. We define:

8.8 $$\xi(\gamma) \equiv \sqrt{k^2 - \gamma} \quad , \quad \eta(\gamma) \equiv \sqrt{n_1^2 k^2 - \gamma} \quad , \quad 0 \leq \gamma \leq k^2$$

Then, the solutions of (8.7) have the form:

8.9 $$\mu(\gamma,z) = \begin{cases} A_\gamma \sin \eta z \quad , & 0 \leq z \leq d \\ A_\gamma[\sin \eta d \cos \xi(z-d) + \frac{\eta}{\xi} \cos \eta d \sin \xi(z-d)], & d \leq z < \infty \end{cases}$$

$$A_\gamma \equiv \sqrt{\xi}\,[\pi(\xi^2 \sin^2 \eta d + \eta^2 \cos^2 \eta d)]^{-1/2} \quad , \quad 0 < \gamma \leq k^2$$

There may, at this point, be some question about solutions (8.9); they do not correspond to intuition. Recall that these solutions are the radiation modes; a superposition of these modes can be used to represent energy lost by propagation into the ocean bottom. One would then quite naturally expect these mode functions to possess an outward, travelling wave structure in the region $d \leq z < \infty$. Instead, equations (8.9) indicate a standing wave structure in this region. In fact, the radiation modes correspond to a plane wave excitation of the configuration at different incidence angles from $z = \infty$, i.e. from infinitely far in the ocean bottom. The resolution of this difficulty lies in the fact that these modes are not physically meaningful individually. It is only a superposition of these modes that makes physical sense. The situation is somewhat analogous to that encountered in Fourier sine or cosine transform theory where finite energy signals are constructed via a superposition of these infinite energy, standing wave trigonometric functions. A good discussion of this point is given by Marcuse in [16] .

We have alluded to the fact that an angle can be associated with the modes of the problem. This angular parametrization will now be made explicit. We introduce the angle θ by defining the longitudinal and transverse wavenumbers to be:

8.10 $$\beta \equiv n_1 k \cos\theta \quad , \quad \sigma \equiv n_1 k \sin\theta$$

This parametrization is shown schematically in Figure 4. In Figure 4 we have also depicted an angle θ_o which is defined in a natural way by the "continuity of longitudinal wavenumber requirement across the interface. Since $k < \beta < n_1 k$, the angle θ ranges from 0 to a critical angle $\theta_c < \pi/2$ defined by the relation:

8.11 $$n_1 k \cos\theta_c = k \qquad \text{or} \qquad \theta_c = \cos^{-1}(1/n_1)$$

From Figure 4, we observe that $\theta = \theta_c$ corresponds to $\theta_o = 0$. Consequently, modes for which $0 < \theta < \theta_c$ (i.e. $k < \beta < n_1 k$) are the bound or guided modes. The mode functions ν_p can be viewed as a superposition of plane acoustic waves whose angle of incidence θ is too small to permit penetration into the bottom. Instead, these waves suffer total internal reflection at the ocean-bottom interface ($z = d$) . As (8.6) indicates, the acoustic field in the bottom region for these modes is exponentially decaying rather than propagating.

Since we shall ultimately be interested in the behavior of the coupled power equations in the high frequency limit, we shall digress now and examine the approximate form of the guided mode function ν_p (given by (8.6)) in the high frequency case.

Using (8.4), eigenvalue equation (8.5) can be recast as:

8.12 $$\tan \sigma_p d = -\frac{\sigma_p d}{\sqrt{(n_1 kd)^2 \sin^2\theta_c - (\sigma_p d)^2}}$$

From this relation it follows that:

8.13 $$\sigma_{p+1} - \sigma_p \cong \pi/d$$

Moreover, for the lower order modes (i.e. $\sigma_p << n_1 k \sin\theta_c$) we have:

8.14 $$\sigma_p \cong p\pi/d$$

Therefore, we can see from (8.6) that as frequency (or equivalently wave number k) increases, the mode functions of the low order guided modes take on the approximate form:

8.15 $$\nu_p(z) \cong \begin{cases} \frac{\sqrt{2}}{d} \sin \frac{p\pi z}{d} , & 0 \leq z \leq d \\ 0 , & d \leq z \leq \infty . \end{cases}$$

Equation (8.15) is what one would intuitively expect. As frequency increases, the exponentially decaying field becomes more tightly bound to the slab. In the limit, the mode function satisfies a pressure-release boundary condition at z = d also.

Having developed the high frequency configuration of the guided mode functions, we shall now consider the random field $\mu(r,z)$ in greater detail. Thus far, we have assumed that μ has an expected value of zero, is localized in depth to the region $0 \leq z \leq d$ and has wide sense stationary matrix elements (c.f. (1.15) and (1.19)). In section 2 we have also mentioned a mixing hypothesis that must be satisfied by these matrix elements in order that the stochastic analysis apply. This hypothesis is tantamount to assuming that statistical independence is achieved in the asymptotic limit as the range separation becomes infinite. Now, however, we shall be more specific in our assumptions for the model problem since we want to explicitly evaluate the cosine transforms defining a_{rp} and b_r in (4.4) and (4.5). Accordingly, we shall assume that the random field μ has the following correlation function:

$$\langle\mu(r,z)\mu(r',z')\rangle \equiv R(z,z')\,\frac{e^{-|r-r'|/\ell}}{\ell} \tag{8.16}$$

where the support of R is contained within $[0,d]\times[0,d]$ since μ vanishes if $z > d$. This decomposition of R into the product of a transverse, depth-dependent correlation and an exponentially-decaying function of range makes the cosine transform evaluation particularly simple. The assumption is, moreover, not as restrictive as it may, at first glance, appear since we could equally well deal with superpositions of the form $\sum_{m=1}^{M} R_m(z,z')\,\frac{e^{-|r-r'|/\ell_m}}{\ell_m}$. We choose as simple a form as possible to best illustrate the relevant features.

From (1.19) and (8.16) it follows that:

$$\langle\mu^*_{pq}(r)\mu_{pq}(0)\rangle = \frac{(n_1k)^4 e^{-r/\ell}}{4\beta_p\beta_q\ell}\int_0^d\int_0^d R(z,z')\nu_p(z)\nu_q(z)\nu_p(z')\nu_q(z')dzdz' \tag{8.17}$$

$$\langle\mu^*_{p\gamma}(r)\mu_{p\gamma}(0)\rangle = \frac{(n_1k)^4 e^{-r/\ell}}{4\sqrt{\gamma}\beta_p\ell}\int_0^d\int_0^d R(z,z')\nu_p(z)\nu(\gamma,z)\nu_p(z')\nu(\gamma,z')dzdz'$$

where the mode functions $\nu_p(z)$ and $\nu(\gamma,z)$ are defined in (8.6) and (8.9) respectively.

Let us use I_{pq} and $I_{p\gamma}$ to denote the double integrals appearing in (8.17). Then, the coefficients a_{pq} and b_p can be expressed as:

$$a_{pq} = \frac{(n_1k)^4\, I_{pq}}{2\beta_p\beta_q[1+(\beta_p-\beta_q)^2\ell^2]} \tag{8.18}$$

$$b_p = \frac{(n_1 k)^4}{2\beta_p} \int_0^{k^2} \frac{I_{p\gamma}\, d\gamma}{\sqrt{\gamma}[1+(\beta_p-\sqrt{\gamma})^2\ell^2]}$$

Since both $\nu_p(z)$ and $\nu(\gamma,z)$ are sinusoidally-varying functions within the region $0 \leq z \leq d$ (c.f. (8.6) and (8.9)), I_{pq} and $I_{p\gamma}$ involve spectral evaluations of the depth-dependent correlation function. In fact, if we define:

$$S(\alpha,\alpha') \equiv \frac{2}{\pi} \int_0^\infty \int_0^\infty R(z,z')\cos\alpha z \cos\alpha' z' dz dz'$$

then I_{pq} and $I_{p\gamma}$ can be expressed in terms of this two-dimensional cosine transform as follows:

8.19 $$I_{pq} = \frac{\pi}{8} A_p^2 A_q^2 [S(\sigma_p-\sigma_q,\sigma_p-\sigma_q) + S(\sigma_p+\sigma_q,\sigma_p+\sigma_q) - S(\sigma_p-\sigma_q,\sigma_p+\sigma_q) - S(\sigma_p+\sigma_q,\sigma_p-\sigma_q)]$$

$$I_{p\gamma} = \frac{\pi}{8} A_p^2 A_\gamma^2 [S(\sigma_p-\eta,\sigma_p-\eta) + S(\sigma_p+\eta,\sigma_p+\eta) - S(\sigma_p-\eta,\sigma_p+\eta) - S(\sigma_p+\eta,\sigma_p-\eta)]$$

Having obtained this representation, we shall introduce a bandlimiting idealization, i.e. we shall assume that the support of S lies within a finite square of (α,α')-space. Theoretically, this assumption is impossible since we have already assumed that $R(z,z')$ has compact support. However, if the ocean depth d is great enough and the random fluctuation μ does not vary too rapidly as a function of the depth variable, this assumption is a reasonable approximation.

We shall, for definiteness, assume that the support of S lies in the region $[0,3\pi/2d]\times[0,3\pi/2d]$ of wavenumber space. Recall from (8.13) that the transverse wavenumber increment for adjacent modes is $\sigma_{p+1}-\sigma_p \cong \pi/d$. Our compact support assumption, therefore, is tantamount to a "nearest neighbor interaction" assumption. With this assumption, (8.19) reduces to:

$$8.20 \qquad I_{pq} = \begin{cases} \frac{\pi}{8} A_p^2 A_q^2 \, S(\sigma_p-\sigma_q, \sigma_p-\sigma_q) \, , & |p-q| \le 1 \\ 0 \, , \text{ otherwise} \end{cases}$$

$$I_{p\gamma} = \begin{cases} \frac{\pi}{8} A_p^2 A_\gamma^2 \, S(\eta-\sigma_p, \eta-\sigma_p) \, , & |\eta-\sigma_p| \le \frac{3\pi}{2d} \\ 0 \, , \text{ otherwise} \end{cases}$$

Moreover, from (8.13) we have:

$$8.21 \qquad I_{p,p-1} \cong \frac{\pi}{2d^2} \; S(\pi/d, \pi/d) \equiv \frac{\pi}{2d^2} S_o$$

which is independent of the integer p .

As a consequence of (8.18) and (8.20), the only nonzero a_{pq} coefficient is $a_{p,p-1} = a_{p-1,p}$. For brevity, we define:

$$8.22 \qquad a_p = a_{p,p-1} = a_{p-1,p}$$

Then, equations (4.3) become:

$$8.23 \qquad \frac{d}{d\tau} W_1 = - b_1 W_1 + a_2 (W_2 - W_1)$$

$$\frac{d}{d\tau} W_p = - b_p W_p + a_{p+1} (W_{p+1} - W_p) - a_p (W_p - W_{p-1}), \quad 2 \le p \le N-1$$

$$\frac{d}{d\tau} W_N = - b_N W_N - a_N (W_N - W_{N-1})$$

We shall use Δ and δ to denote froward and backward difference operators, respectively, i.e.:

8.24 $$\Delta W_p \equiv W_{p+1} - W_p \quad , \quad \delta W_p \equiv W_p - W_{p-1}$$

Then, the general relation in (8.23) can be expressed as:

8.25 $$\frac{d}{d\tau} W_p = - b_p W_p + \frac{\Delta}{\Delta p}\left[a_p \left(\frac{\delta}{\delta p}(W_o)\right)\right] \quad , \quad 2 \leq p \leq N-1$$

Notice that in (8.25), we have simply rewritten unity as:

8.26 $$\Delta p = (p+1)-p = 1 \quad , \quad \delta p = p-(p-1) = 1$$

From (8.14), we note that the integer p can also be expressed as:

8.27 $$p \cong \frac{d}{\pi}\sigma_p = \frac{n_1 kd}{\pi} \sin\theta_p$$

For any given value of p, approximation (8.27) tends to equality as the wavenumber k becomes infinite. From (8.18) and (8.21) we have:

8.28 $$a_p = \frac{(n_1 k)^2 \ (\pi/8)\ A_p^2 A_{p-1}^2 S(\sigma_p - \sigma_{p-1}, \sigma_p - \sigma_{p-1})}{2\cos\theta_p \cos\theta_{p-1}[1+(n_1 k\ell)^2(\cos\theta_p - \cos\theta_{p-1})^2]}$$

Equation (8.27) implies that $\theta_p - \theta_{p-1} = O((n_1 kd)^{-1})$ as $k \uparrow \infty$. Therefore, noting (8.15), it follows that:

8.29 $$a_p = (n_1 k)^2 \ \frac{\pi S_o}{4d^2\cos^2\theta_p[1+(\pi\ell/2d)^2\sin^2\theta_p]} + O(1)$$

as $k \uparrow \infty$. Therefore, if we use representation (8.27) for the finite differences Δp and δp, i.e.

$$8.30 \qquad \Delta p = \frac{n_1 kd}{\pi}(\sin\theta_{p+1} - \sin\theta_p) \quad , \quad \delta p = \frac{n_1 kd}{\pi}(\sin\theta_p - \sin\theta_{p-1})$$

in (8.25) we obtain a net factor $(n_1 k)^2$ in the denominator which is balanced by the $(n_1 k)^2$- dependence of a_p .

These observations point the way to the desired diffusion approximation. We shall define a continuous variable x as:

$$8.31 \qquad x \equiv \frac{d}{\pi}\sin\theta \quad , \quad 0 \le x \le \frac{d}{\pi}\sin\theta_c \equiv x_c < \frac{d}{\pi}$$

and view the coupled power equations as finite difference approximations to a partial differential equation of diffusion type defined in terms of independent variables τ and x . Thus, we interpret $W_p(\tau)$, b_p and a_p as sampled values of functions which we shall denote by $W(\tau,x)$, $b(x)$ and $a(x)$, respectively. We can see from equation (8.30) that this interpretation is appropriate. As $k \uparrow \infty$, the number of guided modes, N, becomes infinite and the set of sampling points $\{x_p\}_{p=1}$ becomes dense in $[0,x_c]$. We obtain the correspondence:

$$8.32 \qquad \frac{\Delta}{\Delta p}\left[a_p \frac{\delta}{\delta_p} W_{pp}(\tau)\right], \quad 2 \le p \le N-1 \quad \longrightarrow \quad \frac{\partial}{\partial x}\left[a(x)\frac{\partial}{\partial x} W(\tau,x)\right], \quad 0 \le x < x_c$$

where

$$8.33 \qquad a(x) \equiv \frac{\pi S_o}{4d^2[1-(\pi x/d)^2][1+(\pi\ell/2d)^2(\pi x/d)^2]}$$

Moreover, if the range correlation length ℓ is much less than the ocean depth d, i.e. $\pi\ell/2d \ll 1$, then:

$$8.34 \qquad a(x) \cong \frac{\pi S_o}{4d^2[1-(\pi x/d)^2]}$$

Lastly, if $\sin\theta_c \ll 1$, i.e. if n_1 is only slightly greater than one, then:

$$8.35 \qquad a(x) \cong \frac{\pi S_o}{4d^2} = a(0) \equiv a_o$$

We now consider the radiation loss coefficient b_p for $2 \leq p \leq N-1$. As (8.20) indicates $I_{p\gamma}$ is proportional to $S(\eta-\sigma_p,\eta-\sigma_p)$. However, from (8.8) and the fact that $0 \leq \gamma \leq k^2$, it follows that:

$$8.36 \qquad \eta-\sigma_p \geq \sqrt{(n_1k)^2-k^2} = n_1k \sin\theta_c - \sigma_p = n_1k(\sin\theta_c - \sin\theta_p)$$

Consider now an arbitrary but fixed value of θ, such that $0 < \theta < \theta_c$, and let the index vary appropriately so that $\lim_{k\uparrow\infty} \theta_p = \theta$. From (8.36), since $\sin\theta_c - \sin\theta > 0$, it follows that

$$8.37 \qquad \lim_{k\uparrow\infty} \eta-\sigma_p = \infty$$

However, as a consequence of the band limiting assumption, $S(\eta-\sigma_p,\eta-\sigma_p)$ will be zero whenever $|\eta-\sigma_p| > 3\pi/2d$. Therefore, for that particular value of θ, the limiting value of the radiation loss term, $-b_pW_p$, is zero. Since the value of θ was arbitrary, we conclude that the continuous variable counterpart to (8.25) must be:

$$8.38 \qquad \frac{\partial}{\partial\tau} W(\tau,x) = \frac{\partial}{\partial x}\left[a(x)\frac{\partial}{\partial x} W(\tau,x)\right] , \quad 0 < x < x_c$$

Because of the bandlimiting assumption, coupling to radiation will occur only for modes whose angle θ is essentially the cirtical angle. Energy initially possessed by a mode whose transverse wavenumber is appreciably less than $n_1k \sin\theta_c$ can not be converted directly to radiation loss. Rather, this energy must first diffuse among the guided modes. Only when this energy is finally coupled to a mode near the edge of the band (i.e. $p \cong N$) can this energy subsequently be coupled to the continuous spectrum and be lost. Therefore, as (8.38) indicates, radiation effects are absent from the description of the limiting interior coupling mechanism. As we shall see, however, the presence of radiation loss will play an important role in defining the boundary condition at $x = x_c$.

To obtain the boundary conditions at $x = 0$ and $x = x_c$, we shall examine the finite difference equations for $p = 1$ and $p = N$, respectively. Recall that for $2 \leq p \leq N-1$, the $(n_1k)^2$- dependence of a_p was balanced by a similar growth arising from the Δp and δp terms (c.f. (8.29) and (8.30)). For p=1 and p=N, however, such a balance will not exist. The identification of these equations as finite difference approximants to continuous variable equations will produce terms that become infinite as $k \uparrow \infty$. We shall obtain the boundary conditions by suppressing this growth, i.e. by demanding that the coefficients of these terms that grow with wavenumber be zero.

Let us first consider the equation for p=1 and deduce the boundary condition for $x = 0$. The radiation loss term, $-b_1W_1$, will clearly not play a role. Therefore, from (8.23), we see that we must consider the equation:

$$8.39 \qquad \frac{d}{d\tau} W_1(\tau) = a_2(W_2(\tau) - W_1(\tau)) = a_p \frac{\delta}{\delta_p} W_p(\tau)\Big|_{p=2}$$

From (8.27) and (8.31), however, observe that:

$$8.40 \qquad x_2 = \frac{d}{\pi} \sin\theta_2 = 2/n_1k \downarrow 0 \quad \text{as} \quad k \uparrow \infty .$$

Therefore, the continuous variable counterpart of (8.39) is:

$$8.41 \qquad \frac{\partial}{\partial\tau} W(\tau,0) = (n_1k) \quad a_o \frac{\partial}{\partial x} W(\tau,0) \quad + O(1) \qquad \text{as} \quad k \uparrow \infty$$

Since $a_o \neq 0$, we require that:

$$8.42 \qquad \frac{\partial}{\partial x} W(\tau,0) = 0$$

The boundary condition at $x = x_c$ is obtained in a similar way; the situation is complicated somewhat, however, by the fact that the radiation term now plays a significant role. From (8.23), the equation that we consider is:

8.43 $$\frac{d}{d\tau}W_N(\tau) = -b_N W_N(\tau) - a_N(W_N(\tau) - W_{N-1}(\tau))$$

Observe that:

8.44 $$x_N = \frac{d}{\pi}\sin\theta_N = \frac{N}{n_1 k} \uparrow x_c \equiv \frac{d}{\pi}\sin\theta_c \quad \text{as} \quad k \uparrow \infty$$

Therefore, we obtain the correspondence:

8.45 $$a_N(W_N(\tau) - W_{N-1}(\tau)) = a_p \frac{\delta}{\delta_p} W_p(\tau)\bigg|_{p=N} \rightarrow (n_1 k)\ a(x_N)\frac{\partial}{\partial x}W(\tau,x_N) +$$

$$O(1) \longrightarrow (n_1 k)\ \ a(x_c)\ \frac{\partial}{\partial x}W(\tau,x_c)\ \ + O(1) \quad \text{as} \quad k \uparrow \infty$$

Let us now consider the radiation term, $- b_N W_N(\tau)$. From (8.18) we have:

8.46 $$b_N = \frac{(n_1 k)^4}{2\beta_N} \int_0^{k^2} \frac{I_{N\gamma}\, d\gamma}{\sqrt{\gamma}\ [1+(\beta_N-\sqrt{\gamma})^2 \ell^2]}$$

where $I_{N\gamma}$ is given by (8.20). Using (8.9) we have:

8.47 $$b_N = \frac{(n_1 k)^4}{8\beta_N} A_N^2 \int_{n_1 k\sin\theta_c}^{n_1 k} \frac{\xi\eta S(\eta-\sigma_N,\eta-\sigma_N)\quad d\eta}{\sqrt{\gamma}[1+(\beta_N-\sqrt{\gamma})^2\ell^2][\xi^2\sin^2\eta d+\eta^2\cos^2\eta d]}$$

where ξ and η are defined by (8.8). Because of the bandlimiting assumption, the actual (nonzero) range of integration extends from $n_1 k \sin\theta_c \equiv \eta_c$ to $\eta_c + 3\pi/2d$. We are interested in determining the asymptotic behavior of b_N as $k \uparrow \infty$. Therefore, we observe that for $\eta_c \le \eta \le \eta_c + 3\pi/2d$, the term $(\beta_N-\sqrt{\gamma})^2$ remains $O(1)$ since β_N and $\sqrt{\gamma}$ are both equal to $n_1 k \cos\theta_c + O(1)$. Therefore, to simplify our computations, we shall assume that the spectral density function is flat over the range of integration and equal to the constant S_o (c.f. (8.21)). With these simplifications, (8.47) becomes:

$$8.48 \qquad b_N \sim \frac{(n_1k)^2 S_o}{4d\cos^2\theta_c} \int_{\eta_c}^{\eta_c+3\pi/2d} \frac{\eta\sqrt{\eta^2-\eta_c^2}\; d\eta}{\eta^2-\eta_c^2+\eta_c^2\cos^2\eta d}$$

where we have used the fact that $\xi = \sqrt{\eta^2-\eta_c^2}$ and the approximation $A_N^2 \cong 2/d$. Define $\phi \equiv (\eta-\eta_c)d$. Then, (8.48) becomes:

$$8.49 \qquad b_N \sim \frac{(n_1k)^2 S_o}{4d^2\cos^2\theta_c} \qquad \frac{2}{\eta_c d} \int_0^{3\pi/2} \frac{\sqrt{\phi}\; d\phi}{(\frac{2}{\eta_c d})\;\phi+\cos^2(\phi+\eta_c d)}$$

To exhibit the frequency dependence of the integral it suffices to assume that $\cos(\phi_o + \eta_c d) = 0$ for one value of ϕ_o lying in $(0,3\pi/2)$; the argument for two zeros or endpoint zeros is basically the same. Choosing a small, fixed value of $\delta > 0$, we have:

$$8.50 \qquad \int_0^{\phi_o-\delta} + \int_{\phi_o+\delta}^{3\pi/2} \frac{\sqrt{\phi}\; d\phi}{(\frac{2}{\eta_c d})\phi+\cos^2(\phi+\eta_c d)} < \frac{2}{3}\,\frac{3\pi}{2}^{3/2}\csc^2\delta = O(1) \text{ as } k \uparrow \infty$$

and

$$8.51 \qquad \int_{\phi_o-\delta}^{\phi_o+\delta} \frac{\sqrt{\phi}\; d\phi}{(\frac{2}{\eta_o d})\phi+\cos^2(\phi+\eta_c d)} \sim \sqrt{\phi_o} \int_{-\delta}^{\delta} \frac{d\psi}{(2/\eta_c d)\phi_o\psi^2} =$$

$$= \sqrt{2\eta_c d\phi_o}\,\tan^{-1} \quad \delta\;\frac{\eta_c d}{2} \quad = O(\sqrt{\eta_c d}) = O(\sqrt{\eta_1 k}) \quad \text{as} \quad k \uparrow \infty .$$

From (8.49)-(8.51), we conclude that $b_N \sim (n_1k)^2$ as $k \uparrow \infty$. Since our simplifying assumptions did not qualitatively alter matters, we expect that such an order of growth prevails in the general case also.

For convenience, we define:

8.52 $$b_N \equiv (n_1 k)\, b(k), \text{ where } \lim_{k \uparrow \infty} b(k) = \infty$$

Then, the continuous-variable counterpart of (8.43) becomes:

8.53 $$\frac{\partial}{\partial \tau} W(\tau, x_N) = -(n_1 k)\, [b(k) W(\tau, x_N) + a(x_N) \frac{\partial}{\partial x} W(\tau, x_N)] \quad + O(1)$$

Recall that $x_N \uparrow x_c$ as $k \uparrow \infty$. Therefore, guided by the ansatz of suppressing growth as $k \uparrow \infty$, we obtain the following boundary condition at $x = x_c$:

8.54 $$b(k) W(\tau, x_c) + a(x_c) \frac{\partial}{\partial x} W(\tau, x_c) = 0$$

Thus, we have a frequency-dependent boundary condition of impedance type. This boundary condition is intuitively very appealing. Notice that if radiation effects were not present, i.e. $b(k) = 0$, then the boundary condition would reduce to one of reflecting type, i.e. $\frac{\partial}{\partial x} W(\tau, x_c) = 0$, since $a(x_c) \neq 0$. This condition, together with (8.42) and equation (8.38), would imply a conservation of energy in the continuous variable case since:

8.55 $$\frac{d}{d\tau} \int_0^{x_c} W(\tau, x) dx = \int_0^{x_c} \frac{\partial}{\partial x} \quad a(x) \frac{\partial}{\partial x} W(\tau, x) \quad dx = a(x_c) \frac{\partial}{\partial x} W(\tau, x_c) -$$

$$a(0) \frac{\partial}{\partial x} W(\tau, 0) = 0$$

In the presence of radiation loss, since $b(k) \uparrow \infty$ as $k \uparrow \infty$, we see that boundary condition (8.54) becomes increasingly absorptive as wavenumber increases. In the limit $k = \infty$, (8.54) reduces to the absorptive boundary condition $W(\tau, x_c) = 0$.

In summary, then, we restate the limiting continuous variable diffusion approximation.

8.56 $$\frac{\partial}{\partial\tau} W(\tau,x) = \frac{\partial}{\partial x}\, a(x)\frac{\partial}{\partial x}W(\tau,x), \quad 0 < x < x_c \equiv \frac{d}{\pi}\sin\theta_c$$

$$\frac{\partial}{\partial x} W(\tau,0) = 0 \;, \qquad b(k)W(\tau,x_c) + a(x_c)\frac{\partial}{\partial x} W(\tau,x_c) = 0$$

To complete the specification of initial-boundary value problem (8.56), initial data, i.e. $W(0,x) \equiv W_o(x)$, must be given. We obtain the initial data by viewing the initial conditions of discrete system (8.23), i.e. $\{W_p(0)\}_{p=1}^N$, as sampled values at $x = x_p$ of the limiting function that we have called $W_o(x)$.

Note that if approximation (8.35) is applicable, we obtain the following explicit solution of (8.56):

8.57 $$W(\tau,x) = \sum_{n=1}^{\infty} A_n e^{-a_o\lambda_n^2\tau} \cos\lambda_n x$$

where the eigenvalue equation is:

8.58 $$b(k)\cos\lambda_n x_c = \lambda_n a(x_c)\sin\lambda_n x_c \;, \qquad n=1,2,\ldots$$

and the coefficient A_n is given by:

8.59 $$A_n \equiv \frac{\int_0^{x_c} W_o(x)\cos\lambda_n x\,dx}{\int_0^{x_c} \cos^2\lambda_n x\,dx} \;, \qquad n=1,2,\ldots$$

Appendix A. Numerical study

In this section we present the results of a numerical study conducted to compare the coupled power equations of section 4 with the high frequency diffusion model developed in section 8. A comparison is made in the two cases where radiation loss is both absent and present.

For simplicity, we have adopted the assumptions that lead to approximations (8.35). Consequently, we assume a nearest-neighbor interaction among the bound modes with $a_{p,p-1} \equiv a_p = (n_1 k)^2 a_o$; $p=1,\ldots,N$. Also, radiation loss, when present, is assumed to occur through coupling between the Nth bound mode and the continuous spectrum; thus, $b_p = 0$, $1 \leq p \leq N-1$ and $b_N \equiv (n_1 k)^2 \beta a_c$ where β is some positive constant. Therefore, if we define the independent variable $\xi \equiv (n_1 k)^2 a_c \tau$, the coupled power equations become:

8A.1
$$\frac{d}{d\xi} W_1 = W_2 - W_1$$

$$\frac{d}{d\xi} W_p = W_{p+1} - 2W_p + W_{p-1} \;, \quad 2 \leq p \leq N-1$$

$$\frac{d}{d\xi} W_N = -(\beta+1) W_N + W_{N-1}$$

For the case of no radiation loss, we simply set $\beta = 0$.

We compare discrete system (8A.1) with the solutions of the diffusion equation:

8A.2
$$\frac{\partial}{\partial \tau} W(\tau,x) = a_o \frac{\partial^2}{\partial x^2} W(\tau,x) \;, \qquad \tau \geq 0 \;; \qquad 0 \leq x \leq x_c$$

subject to the boundary conditions:

8A.3a
$$\frac{\partial}{\partial x} W(\tau,0) = 0 \;, \qquad \frac{\partial}{\partial x} W(\tau,x_c) = 0 \qquad \text{(no radiation loss)}$$

8A.3b
$$\frac{\partial}{\partial x} W(\tau,0) = 0 \;, \quad W(\tau,x_c) = 0 \qquad \text{(radiation loss)}$$

Observe that, for simplicity, we have adopted the limiting absorptive boundary condition for the case where radiation loss is present. Let $y \equiv x/x_c$ and recall that $x_c \cong N/n_1 k)$ (c.f. (8.27) and (8.31)). The solutions to (8A.2) with boundary conditions (8A.3a) and (8A.3b) can be expressed as:

8A.4a $$W = \sum_{n=0}^{\infty} A_n e^{-(n\pi/N)^2\xi} \cos n\pi y \;, \qquad \text{(no radiation loss)}$$

8A.4b $$W = \sum_{n=0}^{\infty} B_n e^{-((n+1/2)\pi/N)^2\xi} \cos(n+1/2)\pi y \;, \qquad \text{(radiation loss)}$$

where A_n and B_n represent the appropriate Fourier coefficients and $0 \leq y \leq 1$. (For simplicity, we shall use the notation $W(\xi,y)$.).

The initial conditions that were adopted correspond to an initial excitation of the lowest order mode; thus, for (8A.1) we assumed that:

8A.5 $$W_1(0) = 1 \;; \qquad W_p(0) = 0 \;, \quad p=2,\ldots,N$$

For the diffusion approximation (8A.2), (8A.3) we have adopted initial conditions

8A.6 $$W(0,y) = \delta(y)$$

so that $\int_0^1 W(0,y)dy = \sum_{p=1}^{N} W_p(0) = 1$. With that choice of initial condition, the Fourier coefficients in (8A.4) become:

8A.7 $$A_o = 1 \;, \quad A_n = 2 \;, \; n = 1,2,\ldots \;; \quad B_n = 2 \;, \quad n = 0,1,2,\ldots$$

The numerical study was conducted for a 10 mode case, i.e. N = 10 . Equations (8A.1), with initial condition (8A.), were integrated numerically to $\xi = 50$, with β set equal to 0 and 1 in the respective cases of no radiation and radiation loss. These results were compared with the diffusion approximation solutions, where the infinite series was found to be adequately approximated by the sum of the first 5 terms, (i.e. $n = 0,\ldots,4$). Moreover, we have used the approximation:

8A.8 $$\int_{(p-1)/N}^{p/N} W(\xi,(p-1/2)/N) \cong \frac{1}{N} W(\xi,(p-1/2)/N)$$

Figure 3. Modal Power vs. Normalized Transverse Wavenumber; 10 Mode Case.

The points described under captions (i) below represent plots of modal power computed using the coupled power equations. The results obtained using a separation of variables solution of the diffusion approximation are described under captions (ii); these latter points are connected by dashed and solid line segments. Note that the coupled power equations and the diffusion approximation generate virtually indistinguishable results.

No Radiation Loss; 10 Mode Case

(a) $\xi=0.5$:

i) $W_p(.5)$, $1 \leq p \leq 10$: ● ● ●

ii) 0.1 $W(.5,(p-.5)/10)$, $1 \leq p \leq 10$, connected by: —— – —— –

(b) $\xi=5.0$:

i) $W_p(5)$, $1 \leq p \leq 10$: o o o

ii) 0.1 $W(5,(p-.5)/10)$, $1 \leq p \leq 10$, connected by: — — — — —

(c) $\xi=50.0$:

i) $W_p(50)$, 1 _ p _ 10: △ △ △

ii) 0.1 $W(50,(p-.5)/10)$, $1 \leq p \leq 10$, connected by: – – – – – –

Radiation Loss; 10 Mode Case; $\beta = 1$

(d) $\xi=50.0$

i) $W_p(50)$, $1 \leq p \leq 10$: □ □ □

ii) 0.1 $W(50,(p-.5)/10)$, $1 \leq p \leq 10$, connected by: ————————

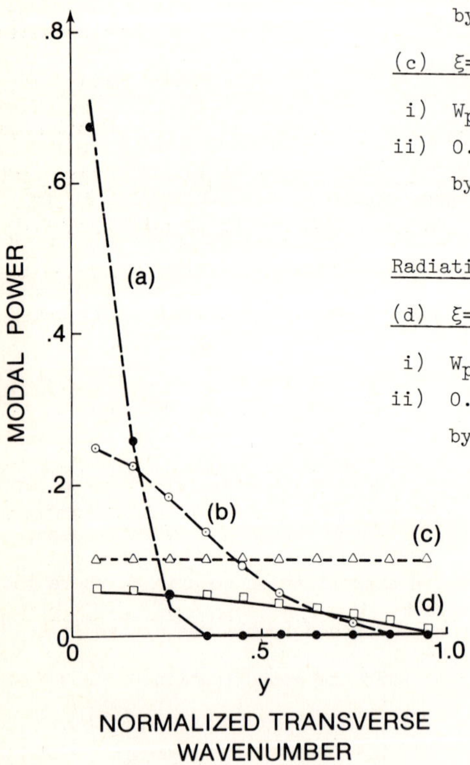

in the comparison of the diffusion approximation solutions with the modal powers, $W_p(\xi)$, $p=1,\ldots,N$.

The results are presented in Figure 3. In this figure, the values of $N^{-1}W(\xi, (p-1/2)/N)$ are plotted at ordinate values $y_p \equiv (p-1/2)/N$, $p=1,\ldots,N$ and these values are linearly interpolated by dashed and solid lines. The values of $W_p(\xi)$, $p=1,\ldots,N$, are also plotted at ordinate values y_p. The values of ξ used are .5, 5 and 50. For the case of radiation loss, only the $\xi = 50$ results are plotted since the data for $\xi = .5$ and 5 essentially coincides with the results presented for the no radiation case.

For the case considered, the continuous variable approximation developed in section 8 provides a good approximation to the system of coupled power equations (8A.1). As ξ increases, the highly-peaked initial power distribution flattens out as energy diffuses into the higher order (initially unexcited) modes. Until an appreciable amount of power becomes coupled into the Nth mode, the effect of radiation loss is negligible. Therefore, the data for $\xi = .5$ and 5 is insensitive to the presence of the radiation loss term. At $\xi = 50$, however, there is a substantial difference between the two cases. In the absence of radiation loss, the power distribution has essentially reached the limiting equipartitioned state. In the presence of radiation loss, the power distribution tapers to 0 as the band edge is approached. There is also a substantial reduction in the total power due to radiation loss. A comparison of curves (c) and (d) indicates that at $\xi = 50$ more than half the initial energy has been radiated.

In Figure 4 we show the results of numerically integrating (6.3) with $N = 10$ under the same hypotheses introduced above. We do not have, at present, a high frequency approximation for second moments of modal powers as we do for first moments.

Appendix B. Diffusion approximation for coupled power equations with radiation loss

In this Appendix we shall analyze the system of equations

8B.1 $$\frac{dW_o^N(\tau)}{d\tau} = N^2(W_1^N(\tau) - W_o^N(\tau))$$

Figure 4. Second Moments of the Modal Powers vs. Normalized Transverse Wavenumber; 10 Mode Case.

In this figure only results corresponding to the coupled fluctuation equations are plotted. The points are connected by dashed and solid line segments simply to facilitate interpretation. No analog of the diffusion approximation has been developed for the second moments.

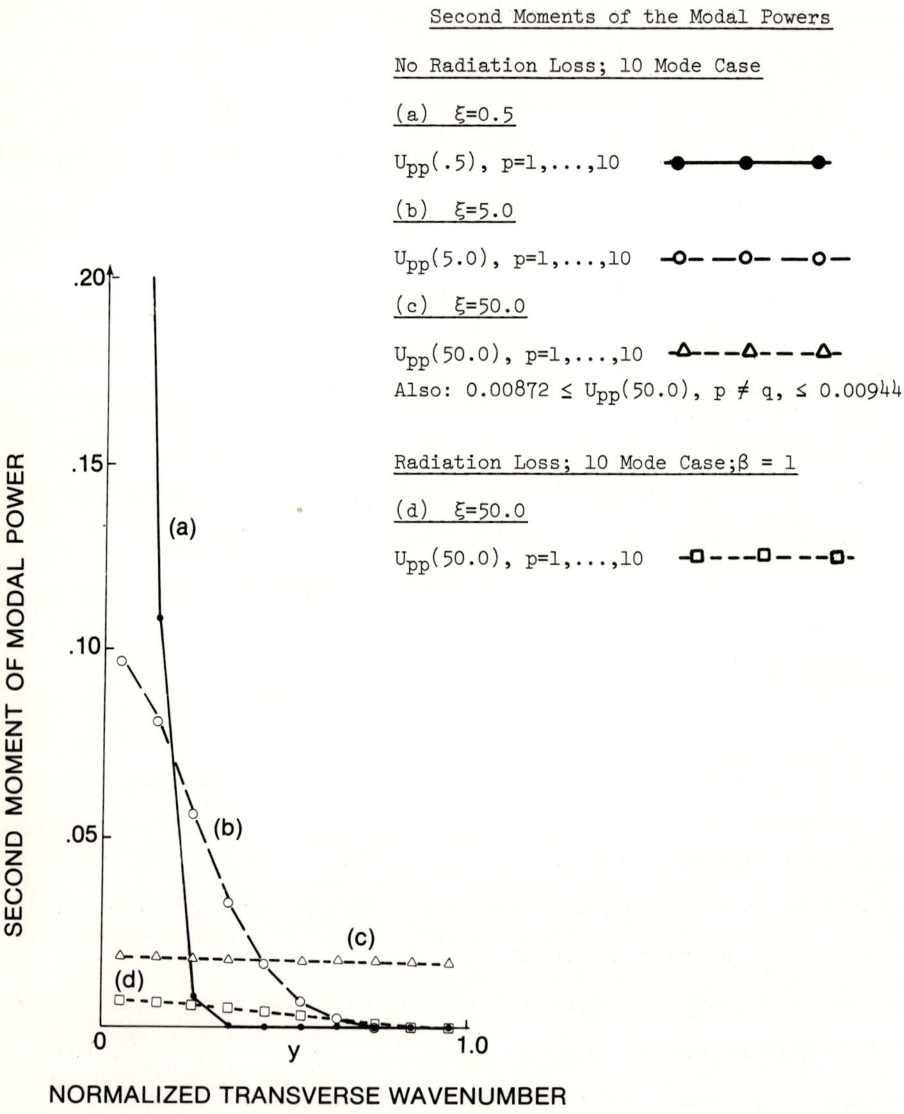

$$\frac{dW_p^N(\tau)}{d\tau} = \frac{1}{2} N^2(W_{p+1}^N(\tau) - 2W_p^N(\tau) + W_{p-1}^N(\tau)) , \qquad 1 \le p \le N-1 ,$$

$$\frac{dW_N^N(\tau)}{d\tau} = N^2(-(1+\beta)W_N^N(\tau) + W_{N-1}^N(\tau)) , \qquad \tau > 0 ,$$

$$W_p^N(0) = f(\tfrac{p}{N}) , \qquad p = 0,1,2,\ldots,N ,$$

where $f(x)$, $0 \le x \le 1$, is a smooth function with compact support in $[0,1]$, in the limit as $N \to \infty$. We shall show that $W_p^N(\tau)$ behaves, asymptotically for $N \uparrow \infty$, like $u^N(\tau,\frac{p}{N})$, $p = 0,1,2,\ldots,N$, where,

8B.2 $$\frac{\partial u^N}{\partial \tau}(\tau,x) = \frac{1}{2}\frac{\partial^2 u^N(\tau,x)}{\partial x^2} , \qquad 0 < x < 1 , \qquad \tau > 0 ,$$

$$u^N(0,x) = f(x)$$

$$\frac{\partial u^N}{\partial x}(\tau,0) = 0 , \qquad \beta u^N(\tau,1) + \frac{1}{N}\frac{\partial u^N}{\partial x}(\tau,1) = 0 .$$

Note that, in turn, $u^N(\tau,x)$ of (8B.2) behaves asymptotically like $u(\tau,x)$ which is the solution of (8B.2) except that

8B.3 $$u(\tau,1) = 0 ,$$

instead of the N-dependent impedance boundary condition. As explained in section 8, it is preferable to work with (8B.2) because it shows dependence on the parameter β and, also, provides a better approximation.

We now proceed with the demonstration of the asymptotic approximation of (8B.1) by (8B.2). First we note that $u^N(\tau,x)$ is a smooth function of $x \in [0,1]$, since the data is smooth, and derivatives are bounded independently of N. Let

8B.4 $$U_p^N(\tau) = W_p^N(\tau) - u^N(\tau,\tfrac{p}{N}) \ .$$

From (8B.1) it follows that

8B.5 $$\frac{dU_o^N(\tau)}{d\tau} - N^2(U_1^N(\tau) - U_o^N(\tau)) = \frac{\partial u^N(\tau,0)}{\partial\tau} - N^2(u^N(\tau,\tfrac{1}{N}) - u^N(\tau,0))$$

8B.6 $$\frac{dU_p^N(\tau)}{d\tau} - \frac{1}{2} N^2(U_{p+1}^N(\tau) - 2U_p^N(\tau) + U_{p-1}^N(\tau))$$

$$= \frac{\partial u^N(\tau,\frac{p}{N})}{\partial\tau} - \frac{1}{2} N^2(u^N(\tau,\tfrac{p+1}{N}) - 2u^N(\tau,\tfrac{p}{N}) + u^N(\tau,\tfrac{p-1}{N}) \ , \quad 1 \leq p \leq N-1 \ ,$$

8B.7 $$\frac{dU_N^N(\tau)}{d\tau} - N^2(-(1+\beta)U_N^N(\tau) + U_{N-1}^N(\tau))$$

$$= \frac{\partial u^N(\tau,1)}{\partial\tau} - N^2(-(1+\beta)u^N(\tau,1) + u^N(\tau,\tfrac{N-1}{N})) \ , \qquad \tau > 0 \ ,$$

$$U_p^N(0) = 0 \ , \qquad p = 0,1,2,\ldots,N \ .$$

We shall show that the right hand sides of (8B.5), (8B.6) and (8B.7) are $0(\frac{1}{N})$. Since τ is in a finite interval, the maximum principle† for (8B.1) tells us that $U_p^N(\tau) = 0(\frac{1}{N})$, $p=0,1,2,\ldots,N$ and the result follows.

Expanding the right hand side of (8B.5) we have

† Equation (8B.1) is, of course, the backward equation for a random walk on the positive integers with boundary condition.

$$N^2(u^N(\tau,\tfrac{1}{N}) - u^N(\tau,0)) - \frac{\partial u^N}{\partial \tau}(\tau,0)$$

$$= N^2(u_x^N(\tau,0)\frac{1}{N} + \frac{1}{2}u_{xx}^N(\tau,0)\frac{1}{N^2} + O(\frac{1}{N^3})) - \frac{\partial u^N}{\partial \tau}(\tau,0)$$

$$= O(\tfrac{1}{N}) ,$$

where we use the boundary condition and the equation (8B.2). Similarly, expanding the right hand side of (8B.6) and using the equation (8B.2) we find that it is also $O(\frac{1}{N})$. Finally, expanding the right hand side of (8B.7) we obtain

$$N^2(-(1+\beta)u^N(\tau,1) + u^N(\tau,1) - u_x^N(\tau,1)\frac{1}{N} + \frac{1}{2}u_{xx}^N(\tau,1)\frac{1}{N^2})$$

$$- \frac{\partial u^N(\tau,1)}{\partial \tau}$$

$$= - N^2(\beta u^N(\tau,1) + \frac{1}{N}u_x^N(\tau,1)) + \frac{1}{2}u_{xx}^N(\tau,1) - \frac{\partial u^N(\tau,1)}{\partial \tau} + O(\frac{1}{N})$$

$$= O(\tfrac{1}{N}) .$$

Here again we use the boundary condition at $x = 1$ and the equation (8B.2).

We have then shown that for any finite τ (fixed)

$$|W_p^N(\tau) - u^N(\tau,\tfrac{p}{N})| = O(\tfrac{1}{N}) , \qquad p = 0,1,\ldots,N , \tag{8B.8}$$

as was intended.

Let us remark that (8B.1) and (8B.2) have been scaled a bit differently than the coupled power equations of section 8 (cf. 8A.1). The differences are not essential however and can be eliminated by changes of variables. In their form (8B.1), the coupled power equations admit the error estimate (8B.8) which is best possible.

References

[1] R. Burridge and H. Weinberg, Chapter III, this volume.

[2] C. Wilcox, Spectral analysis of the Laplacian with a discontinuous coefficient in a half space and the associated eigenfunction expansion. Also, Transient acoustic wave propagation in a symmetric Epstein duct. University of Utah reports, January 1973 and May 1974.

[3] W. H. Munk, Sound channel in an exponentially stratified ocean, with application to SOFAR, J. Acoust. Soc. Am. 55, (1974), p. 220.

[4] F. D. Tappert, Chapter V, this volume.

[5] U. Frisch, Wave propagation in random media, Probabilistic methods in Applied Math., Vol. 1, Academic Press, New York, 1968, pp. 75-198.

[6] R. Kubo, Stochastic Liouville equation, J. Math. Phys. 4 (1963), p. 174.

[7] J. B. Keller, Wave propagation in random media, in Stochastic Processes in Mathematical Physics and Engineering, R. Bellman, editor, Proc. Symp. Appl. Math. XVI (1964), p. 145.

[8] M. Lax, Classical noise IV. Langevin methods, Rev. Modern Phys. 38 (1966), p. 561.

[9] F. Haake, Statistical treatment of open systems by Generalized Master Equations, Springer Tracts in Modern Physics, No. 66, Berlin, Heidelberg, New York, 1973.

[10] G. C. Papanicolaou and W. Kohler, Asymptotic theory of mixing stochastic differential equations, Commun. Pure Appl. Math., 27 (1974), p. 641

[11] R. L. Stratonovich, Topics in the theory of random noise, Vol. 1 and 2, English transl., Gordon and Breach, New York, 1963.

[12] R. Z. Khasminskii, A limit theorem for the solution of differential equations with random right-hand sides, Theor. Probability Appl. 11 (1966), p. 390.

[13] V. Volterra, Theory of Functionals, Dover, New York, 1959.

[14] G. C. Papanicolaou and R. Burridge, Transport equations for the Stokes parameters from Maxwell's equations in a random medium, J. Math. Phys. 16 (1975), p. 2074.

[15] I. I. Gihman and A. V. Skorohod, Stochastic Differential Equations, Springer Verlag, Berlin, Heidelberg, New York, 1972.

[16] D. Marcuse, Theory of Dielectric Optical Waveguides, Academic Press, New York, 1974.

[17] G. C. Papanicolaou and W. Kohler, Asymptotic analysis of deterministic and stochastic equations with rapidly varying components, Commun. Math. Phys., 45 (1975), p. 217.

[18] S. D. Personick, Time dispersion in dielectric waveguides, Bell Syst. Tech. J., 50 (1971), p. 843.

[19] D. Marcuse, Pulse propagation in multimode dielectric waveguides, Bell Syst. Tech. J., 51 (1972), p. 1199.

[20] T. G. Kurtz, A limit theorem for perturbed operator semigroups with applications to random evolutions, J. Funct. Analysis, 12 (1973), p. 55.

[21] G. C. Papanicolaou, Probabilistic problems and methods in singular perturbation, Rocky Mountain Math. J., 6(1976), p. 653.

[22] L. Chernov, Wave propagation in random media, McGraw-Hill, New York, 1960.

[23] H. Weinberg and R. Mellen, Ray diffusion in randomly inhomogeneous ocean, TM No. PA4-63-74, NUSC, New London Laboratory.

[24] R. H. Mellen D.G. Browning and J. M. Ross, Attenuation in randomly inhomogeneous sound chanells, J. Acoust. Soc. Am. 56 (1974), p. 80.

[25] D. Gloge, Optical power flow in multimode fibers, Bell Syst. Tech. J., 51 (1972), p. 1767.

[26] D. Marcuse, Light Transmission Optics, Van Nostrand Reinhold, Princeton, New Jersey, 1972.

[27] L.Dozier, Statistics of Normal Mode Amplitudes in a Random Ocean, Dissertation, N.Y.U., 1977.

CHAPTER V

THE PARABOLIC APPROXIMATION METHOD

Fred D. Tappert

Courant Institute of Mathematical Sciences
New York University
251 Mercer Street
New York, NY 10012

1. Basic concepts

The propagation of acoustic signals in the ocean to long ranges is made possible by the existence of the SOFAR sound channel which acts like a waveguide that confines the acoustic waves within the water column and prevents their interaction with the ocean bottom,which is generally quite lossy compared to the water itself. The parabolic approximation methods discussed in this article are based on the geometrical configurations that naturally arise in the sound channel mode of propagation. By long range propagation we mean propagation to distances of a convergence zone or greater, the convergence zone spacing being about 30 to 35 nmi (1 nmi = 1 nautical mile = 6076.1 ft = 1852 m) or about 50 to 60 km. Since the ocean is about 4 to 5 km deep, we see that sound channel propagation is mainly in a waveguide that is relatively thin vertically and greatly elongated horizontally [31-33]. It is this particular configuration that makes possible the parabolic approximation.

Thus the parabolic approximation in underwater acoustics is quite distinct from the two other main classes of approximations that are commonly used. Geometrical acoustics methods are based on the approximation that wavelengths are small enough so that diffraction effects are negligible everywhere except possibly in a few small regions, and separation of variables methods (such as normal mode expansions) are based on the approximation that the ocean is exactly stratified horizontally so that coupling between the waveguide modes is negligible. Parabolic approximation methods retain all the diffraction effects associated with the particular geometry of the ocean sound channel and thus are valid to much lower frequencies than geometrical acoustics, and they retain the full coupling between waveguide modes and thus are valid for more realistic, non-stratified oceans than separation of variables.

Another important oceanographic fact that is needed in the following discussion is that long-range propagation is necessarily low-frequency, usually below 500 Hz or so. This is because volume absorption of acoustic waves in sea water increases rapidly above about 1000 Hz and because the spectrum of ambient noise often has a broad minimum in the range between 10 and a few hundred Hz. A typical frequency of interest is thus about 150 Hz and the corresponding wavelength is about 10 m. The wavelength is very small compared to the width of the sound channel (about 2 km) and many modes will propagate [31-33].

To make these ideas somewhat more quantitative, let us temporarily adopt the geometrical acoustics and stratified ocean points of view (which are good for making rough estimates) and assume that all bottom interacting rays are attenuated rapidly enough so that they don't contribute to long range propagation. The maximum angle of propagation, also called the "limiting angle", is then given by Snell's law as

$$\theta_\ell = \cos^{-1} (c_{min} / c_{max}) \approx (2\Delta c / c_o)^{1/2} ,$$

where θ is the angle of propagation with respect to horizontal, c_{min} is the minimum sound speed (at the axis of the sound channel), c_{max} is the maximum sound speed (at the bottom of the ocean), $\Delta c = c_{max} - c_{min}$, and c_o is some average sound speed. Typically, $c_o \approx 1500$ m/sec, $\Delta c/c_o \lesssim .04$, and thus $\theta_\ell \lesssim 16^o$. The largest angles of interest in long-range propagation are therefore rather small, and this fact sets the stage for the parabolic approximation.

Actually, most of the energy in sound channel propagation lies within a vertical sector of angles having a half-width of about 5° and this kind of propagation can be viewed as resulting from a sequence of thin lenses as shown in Fig. 1. The effective aperture is $2B \approx 4$ km, the focal length is $R \approx 25$ km, and the focusing angle is $\theta \sim B/R \approx .08$ rad. $\approx 5^\circ$. The f-number of such lenses is large, $f = R/2B \approx 6$, and the Fresnel number, $F = k_o B^2/R = 2\pi B^2/\lambda_o R \approx 100$ at 150 Hz, is also very large.

Although the focusing properties of the ocean sound channel are highly imperfect and full of aberrations, it is clear that an approximation based on weak focusing (large f-number) and the Fresnel theory of diffraction should be adequate.

The basic idea of the continuous Fresnel approximation can be seen in the uniform ocean Green's function expression,

$$p = \frac{1}{[r^2+(z-z_s)^2]^{1/2}} e^{ik_o[r^2+(z-z_s)^2]^{1/2}} \tag{1.1}$$

where z_s is the source depth. If the angle with respect to horizontal is small, i.e.,

$$|\theta| \approx |z-z_s| \;/\; r \ll 1 ,$$

then we may use the approximation

$$p \approx \psi(z,r) \frac{1}{\sqrt{r}} \; e^{ik_o r} , \tag{1.2a}$$

$$\psi(z,r) = \frac{1}{\sqrt{r}} e^{ik_o(z-z_s)^2/2r} \tag{1.2b}$$

It is readily verified that ψ satisfies the parabolic wave equation

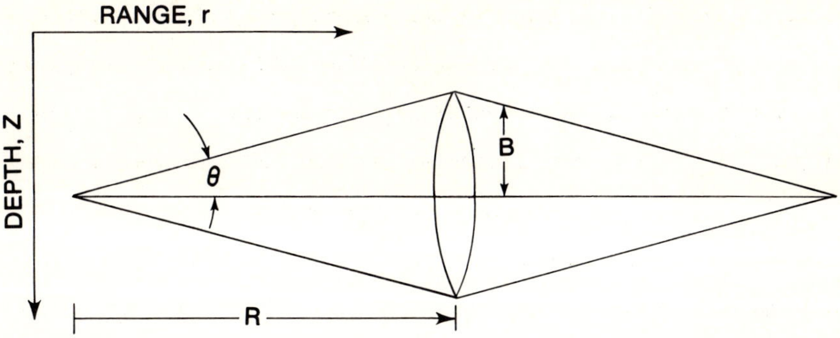

Figure 1. Schematic diagram of weak focusing conditions in the ocean.

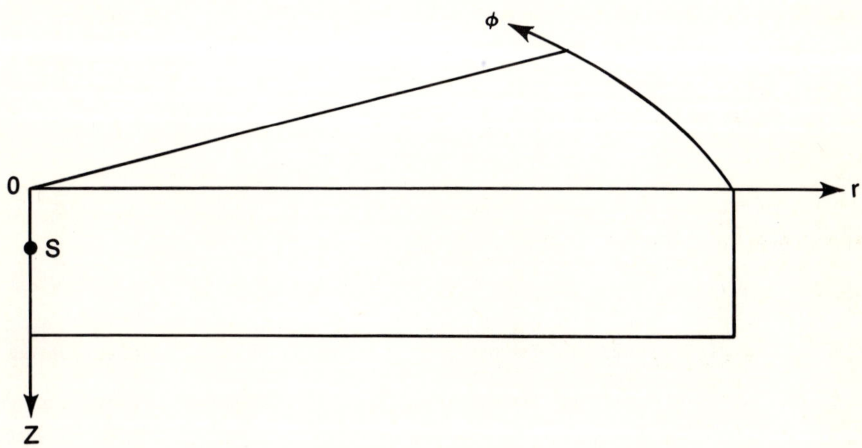

Figure 2. Definition of cylindrical coordinate system.

$$1.3 \qquad i\,\frac{\partial\psi}{\partial r} + \frac{1}{2k_o}\,\frac{\partial^2\psi}{\partial z^2} = 0 \; .$$

This is the most simple example of a parabolic approximation.

Returning to the real ocean problem, elementary physical optics enables one to conclude that in a neighborhood of a focus (convergence zone) the intensity of the acoustic signal will vary significantly over a vertical distance $\Delta z \approx \lambda_o f/\pi$ and a horizontal distance $\Delta r \simeq \lambda_o f^2/\pi$, where $\lambda_o = c_o/\omega = c_o/2\pi\nu$, and ν is the acoustic frequency. Since $f >> 1$, we see that the acoustic field varies very slowly on the scale of a wavelength and this fact suggests that we use an approximation in which the field is represented by a slowly varying envelope with the envelope varying more slowly in range than in depth. Such approximations arise elsewhere in physics and are called parabolic approximations (see Appendix A for an historical discussion).

The most familiar example of a parabolic approximation is that used to describe the slowly varying temporal envelope of a wave packet, or sonar pulse. Denoting the wavenumber as a function of frequency by $k(\omega)$, the integral representation of a wave packet is

$$1.4 \qquad u(r,t) = \int_0^\infty A(\omega)\, \exp[i(k(\omega)r-\omega t)]d\omega \; ,$$

where $A(\omega)$ is the distribution of frequencies. Assuming that this distribution is strongly peaked about the carrier frequency ω_o (narrow bandwidth), we may expand $k(\omega)$,

$$k(\omega) \approx k(\omega_o) + (\omega-\omega_o)\,\frac{dk}{d\omega_o} + \frac{1}{2}(\omega-\omega_o)^2\,\frac{d^2k}{d\omega_o^2} \; ,$$

retaining terms through quadratic (parabolic approximation). (1.2) becomes

1.5a $$u(r,t) \approx \psi(r,t)\exp[i(k(\omega_o)r-\omega_o t)] ,$$

where

1.5b $$\psi(r,t) = \int_0^\infty A(\omega_o+\xi)\exp[i\xi(\frac{dk}{d\omega_o}r-t) + \frac{i\xi^2}{2}\frac{d^2k}{d\omega_o^2}]\,d\xi .$$

Direct differentiation of (1.5) shows that the envelope function ψ satisfies the parabolic equation

1.6 $$i(\frac{\partial\psi}{\partial r} + \frac{1}{v_g}\frac{\partial\psi}{\partial t}) + \frac{\beta}{2}\frac{\partial^2\psi}{\partial t^2} = 0 ,$$

where $v_g = d\omega_o/dk$, $\beta = -d^2k/d\omega_o^2 = v_g^{-3}\,dv_g/dk$. Of course v_g is the group velocity, and β is known as the index of dispersion. As we shall see later, (1.6) also describes the dispersive spreading of acoustic wave packets in a single mode of the ocean sound channel. If the pulse is initially gaussian and given by

1.7 $$\psi(t,o) = p_o\exp(-t^2/2\tau_o^2) ,$$

then at range r the pulse will have width τ given by the relation

1.8 $$\tau^2 = \tau_o^2 + 2\beta^2r^2/\tau_o^2 .$$

This formula will be used later. For now, the main feature to notice is that the parabolic approximation is not concerned with the asymptotic limits $r \to \infty$ or

$\omega_o \to \infty$ as in stationary phase approximations, but is instead concerned with the limit $\Delta\omega/\omega_o \to o$, where $\Delta\omega \sim 1/\tau_o$ is the bandwidth of the function $A(\omega)$. Thus the descriptive phrases, "narrow band approximation", "slowly varying envelope approximation", and "parabolic approximation" are all synonymous.

Returning to the problem of sound channel propagation, we shall make use of a "narrow band of angles approximation" to derive a parabolic equation for the acoustic field by following an analogous procedure. We shall deal with the case of a stratified ocean here to clarify the main ideas, and later in Section 2 will derive the parabolic equations for more realistic oceans. It is well known that the acoustic pressure due to a source of frequency ω can be represented by a sum of propagating normal modes at distance $k_o r \gg 1$ in the form:

1.9 $$p(z,r) = \sum_{m=1}^{M} A_m W_m(z) \frac{1}{(k_m r)^{1/2}} e^{i(k_m r-\omega t)} ,$$

where the W_m are eigenfunctions of the "depth equation",

1.10 $$\frac{d^2W_m}{dz^2} + [k_o^2 n^2(z) - k_m^2]W_m = 0 ,$$

$k_o = \omega/c_o$, $n(z) = c_o/c(z)$, and the k_m are eigenvalues (and radial wavenumbers). The acoustic index of refraction is $n(z)$, the sound speed is $c(z)$, and c_o is some particular value of the sound speed chosen for convenience. Since $n(z)$ differs by only a small amount from unity in the ocean, it is useful to introduce

1.11 $$v(z) = 1-n^2(z)$$

and

$$k_m = k_o(1-\varepsilon_m)^{1/2} .$$

Then (1.10) becomes

1.13 $$\frac{d^2W_m}{dz^2} + k_o^2 [\varepsilon_m - v(z)]W_m = 0 .$$

Since $|v(z)| \lesssim .04$ in the water column, it follows that also $|\varepsilon_m| \lesssim .04$. Thus the allowed values of ε_m lie in a narrow band and we may expand the expression on the right of (1.9). Retaining only the leading term in the exponent and neglecting ε_m in the coefficient yields

1.14 $$p(z,r) \sim \psi(z,r) \frac{1}{(k_o r)^{1/2}} e^{i(k_o r-\omega t)} ,$$

and

1.15 $$\psi(z,r) = \sum_{m=1}^{M} A_m W_m(z) e^{-i\frac{k_o}{2} \varepsilon_m r} .$$

As before, we now differentiate (1.15) with respect to r and use (1.13) to obtain the parabolic wave equation,

1.16 $$i \frac{\partial\psi}{\partial r} + \frac{1}{2k_o} \frac{\partial^2\psi}{\partial z^2} - \frac{k_o}{2} v(z)\psi = 0 .$$

This is the prototype equation of the parabolic approximation in under water acoustics (see Appendix A for further discussion of past and current usage).

Of course in this idealized example the parabolic approximation offers little advantage because the most effective way to solve (1.16) is to separate variables and one then returns to (1.15) and (1.13) and it is rather pointless to make any approximation at all. The real power of the parabolic equation method resides in its ability to handle the more realistic oceans that have horizontal variations of sound speed, water depth, etc. This example does, however, give some insight into the validity of the parabolic approximation in more general cases.

It is clear that the approximation is better, the more narrow the spread of the ε_m. This depends largely on the source excitation functions A_m. If only a single mode is excited, then one can choose c_o such that $\varepsilon_m = 0$ for that mode and there is no error. If only a few modes are excited then one can estimate errors by examining the expansion

1.17 $$e^{ik_m r} = e^{ik_o(1-\varepsilon_m)^{1/2} r} \approx e^{ik_o(1-\frac{1}{2}\varepsilon_m - \frac{1}{8}\varepsilon_m^2)r} .$$

For each term in the sum, the first neglected term in the exponent gives rise to a phase error

1.18 $$\Delta\phi_m = -\frac{k_o}{8}\varepsilon_m^2 r .$$

More important is the relative phase error between two modes that are strongly excited. Using the eigenvalue estimate obtained from a quadratic well of width B, $\varepsilon_m \sim m(\Delta c/c_o)^{1/2}/k_o B$, we find that

$$|\Delta\phi_m - \Delta\phi_{m'}| \sim (m^2 - m'^2)\Delta c/c_o(r/k_o B^2).$$

Thus an optimistic limit of validity is

1.19 $$r \lesssim k_o B^2/(\Delta c/c_o) ,$$

for two adjacent modes. At 100 Hz, using $B \approx 2\text{km}$, $\Delta c/c_o \approx .04$, we obtain

1.20 $$r \lesssim 10^4 \text{km} .$$

More generally, there will be a large number of modes excited and one cannot expect the parabolic approximation to be pointwise accurate over the large range given by (1.20). Moreover, other neglected oceanographic factors (such as random sound speed fluctuations) will destroy the possibility of pointwise accurate predictions of acoustic fields long before the above limit. Typically, significant random point-to-point fluctuations are observed at ranges of a few convergence zones. Thus the most one can try for is that systematic errors be avoided at large ranges.

Another point is that there are many parabolic approximations that are asymptotically equivalent to (1.16), which we first rewrite as

1.21 $$i \frac{\partial\psi}{\partial r} + \frac{1}{2k_o} \frac{\partial^2\psi}{\partial z^2} + \frac{k_o}{2} [n^2(z)-1]\psi = 0 .$$

Since the derivation of this equation required that $|n^2-1| \ll 1$, it is clear that we may equally well replace $\frac{1}{2}(n^2-1)$ by $n-1$ to obtain

1.22 $$i \frac{\partial\psi}{\partial r} + \frac{1}{2k_o} \frac{\partial^2\psi}{\partial z^2} + k_o[n(z)-1]\psi = 0 .$$

Eq. (1.22) has the advantage over (1.21) that when there is no z dependence (unrealistic)

the phase agrees with the WKB expression. A disadvantage is that the eigenfunctions do not agree with the true eigenfunctions for a stratified ocean, as is the case for (1.21). Other possible equations asymptotically equivalent to (1.21) are obtained by replacing k_o by $k_o n(z)$ in the second term giving

1.23a $$i\frac{\partial\psi}{\partial r} + \frac{1}{2k_o n(z)}\frac{\partial^2\psi}{\partial z^2} + \frac{k_o}{2}[n^2(z)-1]\psi = 0, \quad \text{or}$$

1.23b $$i\frac{\partial\psi}{\partial r} + \frac{1}{2k_o n(z)}\frac{\partial^2\psi}{\partial z^2} + k_o[n(z)-1]\psi = 0.$$

The product $k_o n(z)$ does not depend on the choice of c_o and may possibly be an advantage. However, since n must be close to unity for any of these equations to be a valid approximation, there is no theoretical reason to prefer any one of them over the many other possibilities. In section 3 we shall derive parabolic equations that are genuine improvements, and are not asymptotically equivalent to those given here. In any case, it is clear that there are many parabolic equations that can serve as useful approximations. The common element in all such equations is the slowly varying envelope and narrow band approximation.

Since (1.21) is a wave equation, one can obtain a geometrical acoustics approximation directly from this equation. By comparing to the exact geometrical acoustics equation, one gains further insight into the nature of the parabolic approximation and one especially sees that it is a small angle approximation. For a horizontally stratified ocean, the exact ray equations are

1.24 $$\frac{d^2z}{dr^2} = \frac{1}{s^2}\frac{d}{dz}\left(\frac{1}{2}n^2(z)\right),$$

where $s = n(z)\cos\theta = \text{const.}$, and $dz/dr = \tan\theta$. The constant s is called Snell's invariant.

Of the many ways to derive the corresponding ray equations from (1.21), we shall proceed by writing the envelope ψ in polar form in terms of a real amplitude and phase:

1.25 $$\psi(z,r) = A(z,r)e^{i\phi(z,r)} .$$

Substituting (1.25) into (1.21), equating separately the real and imaginary parts to zero, and defining

1.26 $$\theta(z,r) = \frac{1}{k_o}\frac{\partial\phi}{\partial z} ,$$

yields the pair of equations:

1.27a $$\frac{\partial A^2}{\partial r} + \frac{\partial}{\partial z}(\theta A^2) = 0 ,$$

1.27b $$\frac{\partial\theta}{\partial r} + \theta\frac{\partial\theta}{\partial z} = \frac{\partial}{\partial z}\left(\frac{1}{2}n^2(z)\right) + \frac{1}{2k_o^2}\frac{\partial}{\partial z}\left(\frac{1}{A}\frac{\partial^2 A}{\partial z^2}\right) .$$

Eqs. (1.27) are exactly equivalent to (1.21), the last term on the right of (1.27b) describing diffractive effects. Geometrical acoustics is obtained by taking the formal limit $k_o \to \infty$, in which case this last term drops out. Eq. (1.27a) then states that the acoustic power A^2 is transported along the characteristics,

1.28a $$\frac{dz}{dr} = \theta .$$

Thus θ is the (small) angle with respect to horizontal. Eq. (1.27b) becomes

1.28b $$\frac{d\theta}{dr} = \frac{d}{dz}\left(\frac{1}{2}n^2(z)\right) .$$

Combining these two equations gives an equation of the same form as (1.24) except that

1.29 $$s \approx 1 .$$

In order that s be near unity we need that $n \approx 1$ and $\theta \ll 1$ (recall that $s = n \cos\theta$). This gives a succinct statement of the conditions for the validity of the parabolic approximation, provided the acoustic frequency is high enough to justify the use of ray equations. It may also be noted that for any specified ray, (1.28) can be made exactly equivalent to (1.24) by a simple rescaling of the range variable: $r = r'/s$. However, this scaling depends on the angle of emission of the ray so one cannot uniformly rescale an entire family of rays. Thus the small angle condition is still necessary.

The above geometrical acoustic analysis together with the preceeding normal mode analysis shows the main reason why the development of small angle, or parabolic approximation methods arrived so late in the history of underwater acoustics. Namely, for stratified oceans this approximation has very little to offer because it does not lead to any significant simplification. The ray equations in the small angle approximation have the same form as the exact equations, and the normal mode equations in the small angle approximation require the solution of the same eigenvalue problem as the exact equations. For range dependent environments, the situation is quite different. One no longer has Snell's invariant, and instead of (1.24) the exact ray equations take a much more complicated form with several direction cosines that must be recomputed along each ray. In the small angle approximation, however, one obtains a simple generalization of (1.28):

1.30 $$\frac{d^2z}{dr^2} = \frac{\partial}{\partial z}\left[\frac{1}{2}n^2(z,r)\right] ,$$

where now the index of refraction n depends on both z and r . Eq. (1.30) should be useful in numerical ray tracing studies just as the corresponding parabolic wave equation has already proven its utility.

Finally, it may be worthwhile noticing that the geometrical acoustics approximation, just like the parabolic approximation, has errors that accumulate with range. To estimate this error, let us neglect the other error associated with the small angle approximation and examine the error due solely to neglect of diffraction. The relative magnitude of the neglected term in (1.27b) is

1.31 $$\frac{1}{k_o^2 A}\frac{\partial^2 A}{\partial z^2} \sim \frac{1}{k_o^2(\Delta z)^2} \sim \frac{1}{F^2} ,$$

where we have used the previously obtained estimate of the vertical scale of amplitude changes near convergence zones, and F is the Fresnel number, $F = k_o B^2/R$. In one convergence zone period, the relative error in ray position is thus of order $F^{-2} \sim R^2/k_o^2 B^4$, or about 10^{-4} at 150 Hz and 10^{-2} at 15 Hz. This error also shows up as a displacement of the focal plane of an ideal thin lens. Using the uniform medium parabolic equation,

1.32 $$i\frac{\partial\psi}{\partial r} + \frac{1}{2k_o}\frac{\partial^2\psi}{\partial z^2} = 0 .$$

with initial condition

1.33 $$\psi(z,o) = e^{-(z-z_s)^2/2B^2} e^{-ik_o(z-z_s)^2/2R} ,$$

one easily calculates that the distance to the focal plane is $R(1+R^2/k_o^2B^4)^{-1}$ which is slightly less than the distance predicted by geometrical acoustics. Thus ray-trace predictions of the location of convergence zones are always slightly in error, the error becoming larger the greater the range and the lower the frequency. Parabolic approximations are therefore not unique in having errors that accumulate with range.

The main features of the SOFAR sound channel that make possible long-range over-the-horizon ocean acoustic propagation have been outlined. Within a horizontal distance from a low-frequency source equal to a few ocean depths, the acoustic energy propagating at angles greater than the bottom limiting angle is stripped away by lossy bottom interactions leaving only trapped waves propagating at small angles with respect to horizontal and giving rise to predominantly cyclindrical spreading. The natural approximation appropriate to this particular geometrical configuration was shown to be the small angle, narrow band, parabolic approximation. The discussion was centered around stratified oceans, and (1.21) was shown to be the fundamental equation of the parabolic approximation method in this case. In the next section we shall derive a number of more general parabolic wave equations that extend the method to a greater variety of oceanic environments.

2. Derivations of Parabolic Equations

The preceding section dealt heuristically with the parabolic equation method in the overly idealized approximation of a horizontally stratified ocean. In this section we shall strengthen the foundations of the parabolic approximation by providing several alternative derivations and shall further extend the method to include horizontal variations of sound speed, volume absorption, ocean depth, as well as azimuthal (oceanic front) effects, several time-dependent effects, and randomly fluctuating ocean effects. The emphasis will consistently be placed on the derivation of approximation model equations. It is understood that the problem of solving these equations for realistic ocean environments rightfully belongs, in this era of high-speed digital computers, to specialists in numerical analysis and computer science. Let it suffice to say that parabolic wave equations have proven to be remarkably well adapted to efficient machine calculation. This is mainly because they belong to the "marching" class of partial differential equations, i.e., algorithms can readily be devised for the solution of equations of the type of (1.3), (1.6), (1.16), (1.21), etc., in which the acoustic field is advanced one step at a time in range using only information about the field at previously computed ranges [58-62]. In addition, one solves for the slowly varying envelope function itself, and thus computations do not have to be done on the scale of wavelength. This makes feasible the numerical solution of underwater acoustic propagation problems that would be quite impossible with present generation computers if one had to solve directly the elliptic reduced wave equation or the hyperbolic acoustic wave equation. Thus the primary motivation behind parabolic approximation methods is to make controlled reliable approximations right at the beginning of the analysis in order to obtain approximate equations which, even though they may not be analytically soluble themselves, are especially well adapted for efficient high-speed machine calculations.

We shall begin with the case of a fixed monopole (point) source radiating a single frequency in an ocean whose acoustic index of refraction depends on the three spatial coordinates but not on time. We actually have in mind, of course, the situation where the temporal variations of the ocean are so slow that we may neglect any changes of sound speed during the time it takes an acoustic signal to propagate from source to receiver. Thus the time t appears in the index of refraction as a parameter, but we shall not explicitly display this dependence. This approximation would apply, for example, to diurnal variations of acoustic velocity. Also, we shall at first neglect variations of the fluid density and later put this effect back into the model as an effective index of refraction. We shall use the cylindrical coordinate system shown in Fig. 2: z is the depth measured downward from the surface, r is the range measured horizontally, and ϕ is the azimuthal angle (bearing) measured from an arbitrary reference direction. The governing equation is then the reduced wave equation for the acoustic pressure p:

2.1a $$\Delta p + k_o^2[n^2(z,r,\phi) + i\nu(z,r,\phi)]p = -4\pi p_o\delta(\vec{x}) ,$$

2.1b $$\Delta p = \frac{1}{r}\frac{\partial}{\partial r} r \frac{\partial p}{\partial r} + \frac{\partial^2 p}{\partial z^2} + \frac{1}{r^2}\frac{\partial^2 p}{\partial \phi^2} ,$$

2.1c $$\delta(\vec{x}) = \delta(z-z_s)\frac{1}{2\pi r}\delta(r) .$$

Also, $k_o = \omega/c_o$, ω is the angular acoustic frequency, $n(z,r,\phi)=c_o/c(z,r,\phi)$, c_o is a normalization sound speed, and $\nu(z,r,\phi)$ is the volume absorption coefficient. The source strength is p_o (the pressure at unit distance) and it is located at $r = 0$ and depth z_s. One of the basic problems in underwater acoustics is to solve this equation for the acoustic field $p(z,r,\phi)$ given the functions n and ν , and subject to boundary conditions at the surface and bottom.

Since, as will be seen, the parabolic approximation does not alter the surface or bottom boundary conditions, we do not need to dwell on this aspect of the problem. The surface boundary condition is the usual "pressure-release" condition $p(\zeta(r,\phi),r,\phi) = 0$, where $\zeta(r,\phi)$ is the displacement of the surface from the mean level $z = 0$. The boundary condition at the bottom is more difficult to specify in mathematical terms because it depends on how much effort one is willing to spend on modeling propagation through the material layers underlying the ocean floor and on how much one believes these effects influence the acoustic signals at the desired range and location. Physically, the bottom often consists of deep layers (many wavelengths thick) of sediment which behaves acoustically like a fluid with sound speed close to that of water but with much greater volume loss. Acoustic signals propagating at very steep angles may penetrate the sediment layers and then propagate through, and possibly be reflected by, layers of soft (limestone) or hard (granite or basalt) rock. In any case, waves that penetrate sufficiently deep into the sub-bottom layers do not return to the water with enough strength to contribute significantly to long-range propagation and should be removed from the calculation. This effect is modeled by making $\nu(z,r,\phi)$ increase rapidly for values of z much greater than the depth of the ocean and then cutting off the calculational domain at a depth where the acoustic field has been reduced to a negligible amplitude.

Since (2.1) is elliptic, we also have to give a boundary condition on some vertical boundary surrounding the source. This too is difficult to specify because the usual outgoing radiation condition does not apply to an ocean with horizontal variations. Here we encounter a feature of the parabolic approximation that did not arise in the discussion of the previous section, namely, that this approximation automatically eliminates backscattering and reverberation. That is, within the parabolic approximation there is no coupling between outward and inward propagating waves so we do not need to be concerned with a boundary condition on a vertical surface. Although this simplifies the formulation of the acoustic model, it also leads to an additional error which one would sometimes like to avoid. We shall show later how a first order correction can be added to the parabolic equation method which allows the calculation of reverberation.

Let us now proceed to derive the parabolic wave equation from (2.1). As discussed in Section 1, the main idea is that to leading order all significant acoustic waves in the ocean at low frequencies are propagating primarily in the horizontal direction away from the source. Thus the acoustic field may be represented as an outgoing Hankel function $H_o^{(1)}(k_o r)$ which is slowly modulated by an envelope function that depends on depth, range, and azimuth:

2.2 $$p(z,r,\phi) = \psi(z,r,\phi)H_o^{(1)}(k_o r) .$$

This is expected to be a good approximation only in the far field of the point source where $k_o r >> 1$ and

2.3 $$H_o^{(1)}(k_o r) \sim (2/i\pi k_o r)^{1/2} e^{ik_o r} .$$

Substituting (2.2) into (2.1) and omitting the source term because (2.2) is not expected to hold in the immediate neighborhood of the source, we obtain without further approximation:

$$\frac{\partial^2\psi}{\partial r^2} + \left[\frac{2}{H_o^{(1)}} \frac{\partial H_o^{(1)}}{\partial r} + \frac{1}{r}\right] \frac{\partial\psi}{\partial r} + \frac{\partial^2\psi}{\partial z^2} + \frac{1}{r^2}\frac{\partial^2\psi}{\partial\phi^2}$$

2.4 $$+ k_o^2 \; [n^2(z,r,\phi) - 1 + i\nu(z,r,\phi)\;]\psi = 0 \;.$$

We now make the far field approximation, $k_o r \gg 1$, and note from (2.3) that

2.5 $$\frac{2}{H_o^{(1)}} \frac{\partial H_o^{(1)}}{\partial r} + \frac{1}{r} = 2ik_o \;\; [1 + O(\frac{1}{k_o^2 r^2})\;] \;.$$

Neglecting the term of order $(k_o r)^{-2}$, we obtain

2.6 $$\frac{\partial^2\psi}{\partial r^2} + 2ik_o \frac{\partial\psi}{\partial r} + \frac{\partial^2\psi}{\partial z^2} + \frac{1}{r^2}\frac{\partial^2\psi}{\partial\phi^2} + k_o^2[n^2 - 1 + i\nu]\psi = 0 \;.$$

We will leave aside for the moment the problem of connecting the solution of this equation to the field near the source (the source modeling problem), and continue with the main approximation needed to obtain the parabolic wave equation. The required step is clearly to neglect the term $\partial^2\psi/\partial r^2$ compared to the term $2ik_o \; \partial\psi/\partial r$. The way has been prepared in Section 1 for this step and later in this section we shall further analyze the nature of the error committed in making this approximation. For now, it is enough to note that if the main radial dependence of the acoustic field is $\exp(ik_o r)$ for some choice of k_o , then the envelope ψ will vary slowly as a function of r on the wavelength scale, i.e., $\partial\psi/\partial r \ll k_o\psi$ and the neglect of $\partial^2\psi/\partial r^2$ is justified. Neglecting this term in (2.6) yields the fundamental equation of the parabolic equation method in underwater acoustics:

2.7 $$2ik_o \frac{\partial\psi}{\partial r} + \frac{\partial^2\psi}{\partial z^2} + \frac{1}{r^2}\frac{\partial^2\psi}{\partial\phi^2} + k_o^2[n^2(z,r,\phi)-1+i\nu(z,r,\phi)]\psi = 0 \ .$$

Once the field is specified at some range r, (2.7) can be solved as an "initial value problem" by advancing the solution outward in range. The boundary conditions in z were discussed earlier and the solution is periodic in the variable ϕ . Thus we have an acoustic model that allows for variations in sound speed and volume absorption in all three dimensions, and also allows for variable surface height and ocean depth. There are a variety of numerical schemes that provide rapid and accurate solutions to parabolic equations of the type (2.7). As discussed in Section 1, the main sound channel will cause the envelope function ψ to vary on the vertical scale $\Delta z \approx \lambda_o f/\pi$ and horizontal scale $\Delta r \approx \lambda_o f^2/\pi$, where $\lambda_o = 2\pi/k_o$ is the nominal acoustic wavelength and $f \approx R/2B \approx 6$ is the typical f-number of the sound channel. How rapidly ψ varies as a function of ϕ will depend on n and ν . Also, rapid variations of n and ν (or the boundary conditions) will induce corresponding variations of ψ. It is almost universally true that oceanic variations are much more gradual in the horizontal coordinates than in the vertical coordinate. Thus the resolution needed to solve (2.7) will be as stated above and is much longer than the wavelength scale.

Other simplifications of (2.7) are also of considerable practical use. Far from the source, the curvature of the cylindrical wavefronts can be neglected and the azimuthal coordinate in (2.7) can be replaced by a locally cartesian coordinate $dy = rd\psi$ to give

2.8 $$2ik_o \frac{\partial\psi}{\partial r} + \frac{\partial^2\psi}{\partial z^2} + \frac{\partial^2\psi}{\partial y^2} + k_o^2[n^2(z,y,r)-1+i\nu(z,y,r)]\psi = 0 \ .$$

This equation is especially useful for calculations with relatively narrow beams, which are of interest in connection with horizontally extended receivers which select only a narrow band of directions, or for sources which have a directional radiation pattern.

The most widely used parabolic wave equation in underwater acoustics results from (2.7) when the azimuthal derivatives of ψ are neglected. This does not mean that the field is assumed to be cylindrically symmetric (which would be absurd for realistic ocean propagation to long ranges), but rather that the variation of the ocean in azimuth is so gradual that we may neglect scattering from one azimuthal direction to another. We then obtain

2.8 $$2ik_o \frac{\partial\psi}{\partial r} + \frac{\partial^2\psi}{\partial z^2} + k_o^2[n^2(z,r)-1+i\nu(z,r)]\psi = 0 \ ,$$

where the dependence of n and ν on ϕ has been suppressed just as we earlier suppressed their actual dependence on time. In practical applications, one will of course choose the functions n and ν (which also contain the information about bathymetry) to correspond to the particular bearing, time of day, season of year, etc., for which one wants to know the acoustic field.

We now turn to the problem of source modeling, i.e., obtaining initial data for (2.8). There are many ways to get initial data, the best being to solve the full elliptic wave equation in a small region containing the source and extending out several wavelengths in r from the source to the region where the parabolic equation becomes valid. If the ocean can be assumed to be exactly stratified near the source, then this solution can be obtained by separation of variables and calculating the normal modes (including the continuous spectrum which can be important near the source). In many applications, however, this procedure is unnecessarily complicated and a much simpler prescription suffices. This is because one only cares about the energy that is injected into the sound channel and propagates to long ranges. If the source is several wavelengths from any boundary, then we know that near the point source the field will be a spherically spreading wave,

2.9 $$p = \frac{p_o}{R} e^{ik_o R} \ ,$$

where $R = [r^2+(z-z_s)^2]^{1/2}$, Assuming that this solution holds out to a range r such

2.10 $$k_o^{-1} << r << B ,$$

where again B is the width of the sound channel (or roughly the scale length of the thermocline), then we know the field at a range where the parabolic approximation is valid but before any significant refraction effects have occured. In this "overlap" region, (2.8) becomes simply

2.11 $$2ik_o \frac{\partial\psi}{\partial r} + \frac{\partial^2\psi}{\partial z^2} = 0$$

and from (2.2) and (2.3) the pressure is

2.12 $$p(r,z) \approx \psi(z,r)\ (2/i\pi k_o r)^{1/2} e^{ik_o r} .$$

Comparing (2.12) to (2.9) and making small angle approximation, $|z-z_s|/r << 1$, to second order, we see that in the region defined by (2.10) we want ψ to have the approximate form

2.13 $$\psi(z,r) \approx p_o(i\pi k_o/2r)^{1/2}\ [1 - \frac{1}{2}\frac{(z-z_s)^2}{r^2}]\ e^{ik_o(z-z_s)^2/2r} .$$

This can be achieved by starting at r = 0 with a source extended vertically and designed to produce the field in (2.13) when $k_o r >> 1$. The simplest such source, and one which does not produce spurious sidelobes, is

2.14 $$\psi(z,o) = Ae^{-(z-z_s)^2/w^2} ,$$

2.14a $$A = p_o i\sqrt{\pi}/w$$

2.14b $$w = \sqrt{2}/k_o = \lambda_o/\pi\sqrt{2}\ .$$

To show this, we solve (2.11) with initial condition (2.14) to get

2.15 $$\psi(z,r) = p_o(i\ k_o/2r)^{1/2}\ s^{-1/4}\ e^{-k_o^2w^2(z-z_s)^2/4r^2s}$$
$$\times\ e^{\ i[k_o(z-z_s)^2/2rs + \frac{\pi}{4} - \frac{1}{2}\tan^{-1}(2r/k_ow^2)]}\quad ,$$

and $s = 1+k_o^2w^4/4r^2$. If $k_ow = O(1)$, then in the far field, $k_or >> 1$, we have $s \approx 1$ and (2.15) becomes

2.16 $$\psi(z,r) = p_o(i\ k_o/2r)^{1/2}\ e^{-k_o^2w^2(z-z_s)^2/4r^2}$$
$$\times\ e^{ik_o(z-z_s)^2/2r}\quad .$$

Comparing this to (2.13), we see that the behavior is correct when $z = z_s$, the phase is correct, and the distribution of intensity in depth will be correct to second order in angle provided w is chosen according to (2.14b). This prescription has been found to work quite well in practice in the sense that near the source in the forward sector of angles, the intensity $|p|^2$ decreases proportional to r^{-2} (and the transmission loss is 66.13 dB re 1 yd at r = 1 nmi), while at greater ranges one observes a general trend toward a r^{-1} decrease.

Recalling that p_o represents the pressure of the point source at unit distance, which we call r_o , the conventional expression for transmission loss is

$$\text{2.17} \qquad TL = 10 \log_{10} [\ |p(z,r)|^2/(p_o/r_o)^2\ ]$$

$$= 10 \log_{10} [\ \frac{2r_o^2}{\pi k_o r p_o^2} |\psi(z,r)|^2\ ]\ .$$

This expression is clearly independent of p_o , and insertion of the short-range field (2.13) shows that TL would vanish at $r = r_o$ if that formula held at such a short range. By convention, in the U. S. we usually take $r_o = 1$ yd.

Although one may always use the principle of linear superposition to compute acoustic fields from sources that are not point radiators by adding coherently many fields with sources as prescribed above, it is more convenient in practice to directly model the distribution of radiated acoustic energy from the actual source. Since these distributions are conventionally specified in terms of the far-field radiation patterns, and this is just what is needed to begin the integration of the parabolic wave equation, we see that the parabolic equation method is readily adapted to handle a general class of sources (such as directional radiators, etc.). It should also be mentioned here that in practical calculations one usually interchanges the source and receiver, making use of the principle of reciprocity. Thus, for example, if a source moves in range with respect to a fixed receiver and if the environment is range-dependent, one would naturally begin the calculation of acoustic fields at the receiver and march out in range toward the source, thereby avoiding the necessity for recomputing the field at each different range to the source. In addition, what was said above about modeling extended sources applies equally well to modeling extended receivers that have directional properties.

As in other physical sciences, the ultimate validation of an underwater acoustic model must depend on comparisons to experimental data. Computer programs based on the "split-step Fourier" algorithm [35-40] have been implemented in several laboratories in order to obtain solutions of (2.8) and make comparisons to experimental data. These tests have shown that the parabolic equation method performs at least as well as other models for low frequency, deep-ocean, long-range acoustic propagation and sometimes does extraordinarily well, especially in strongly range-dependent environments where other models have great difficulty.

Figures 3 and 4 show an example of such a comparison. The data in Fig. 3 was taken by T. Talpey in the North Pacific [63]. The source was a calibrated sinusoidal projector with nominal frequency of 144 Hz towed at a depth of 280 ft. moving along a certain radial bearing from a fixed receiver at a depth of about 4290 ft. The data shows three convergence zone peaks at ranges of 34, 66, and 97 nmi (the last being split into two distinct sub-peaks); and also shows two lower more diffuse peaks at ranges of about 50 and 82 nmi. The diffuse peaks were interpreted as being caused by reflections from the ocean floor. The sound speed profiles and ocean depths were accurately measured along the track, and were used as input (together with assumed volume loss functions) to a computer code that solved (2.8) and calculated the transmission loss according to (2.17). This numerically calculated result is shown in Fig. 4 on the same scale as Fig. 3. The three convergence zone peaks are seen to be accurately predicted both in amplitude and range, as well as the secondary diffuse peaks which were confirmed to be caused by reflections from the multifaceted sloping ocean bottom. Discrepancies do appear in the shadow zone regions, but the transmission is so poor here that it has no practical consequences. The success of the parabolic equation method in this challenging example gives support to its usefulness and reliability in practical underwater acoustic problems.

Of course, a much greater number of such comparisons in a variety of environmental conditions are needed to thoroughly validate this acoustic model and to determine its range of validity. Admittedly, there are circumstances where this acoustic model has not performed as well as in the above example. For instance, the prediction of transmission via bottom reflected paths that do not lie in a vertical plane (due to tilted bottoms) clearly requires a three dimensional calculation.

Another successful parabolic equation calculation is shown in Fig. 5, which is also concerned with acoustic propagation in the North Pacific Ocean. Along a meridian at about 170° W, the east-flowing Kuroshio Current (about 42° N) marks a separation between cold sub-arctic water and a more temperate body of water characterized by a double thermocline. Propagation across this oceanic front,which is about 625 nmi south from an assumed receiver,results in an enormous (40 dB) decrease in transmission,

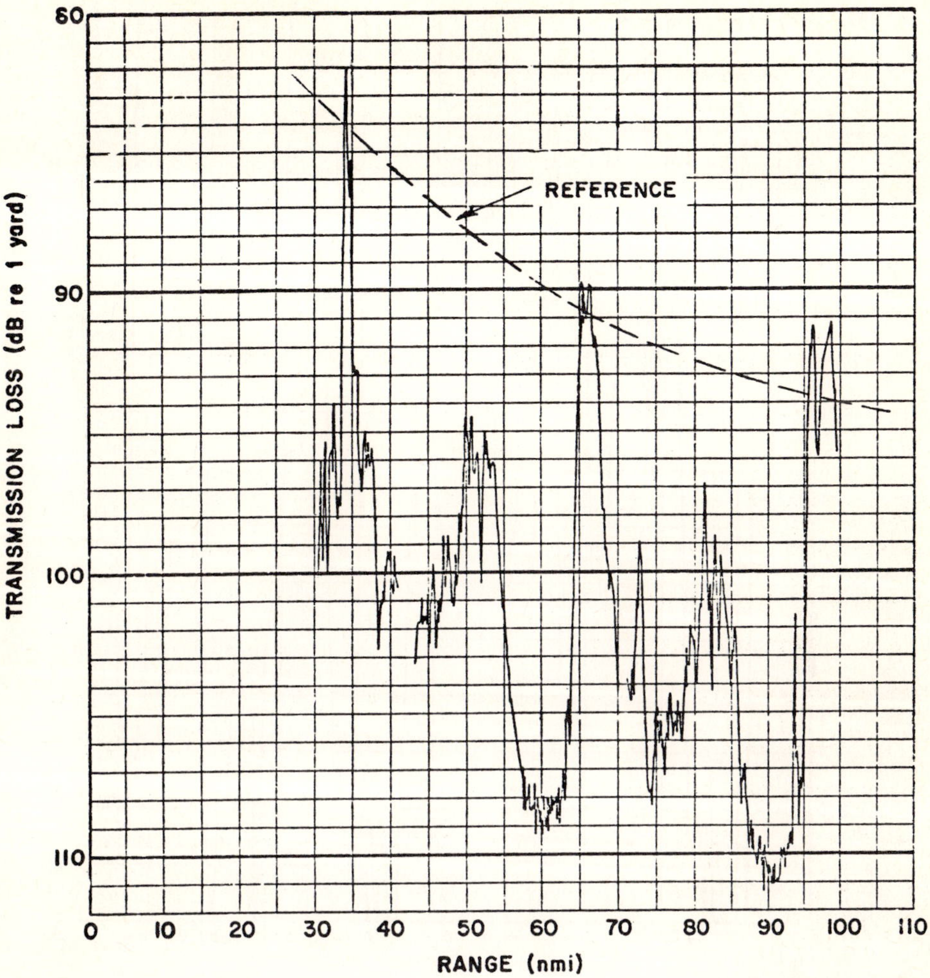

Figure 3. Measured transmission loss (by T. Talpay).

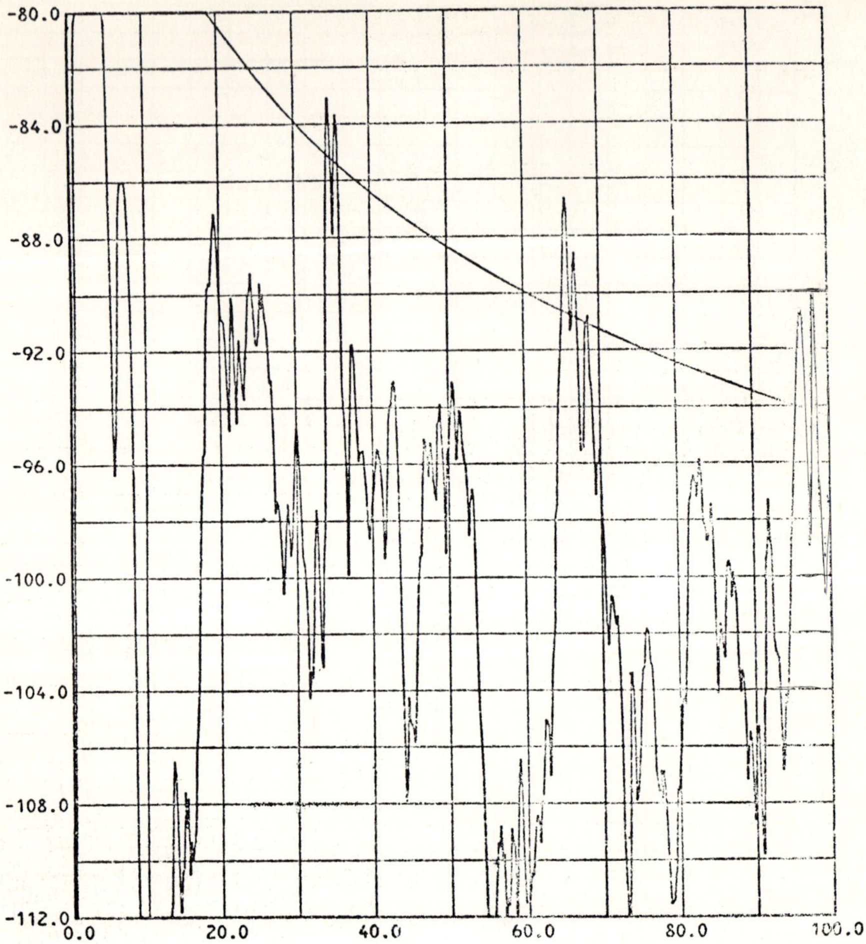

Figure 4. Predicted transmission loss using parabolic equation.

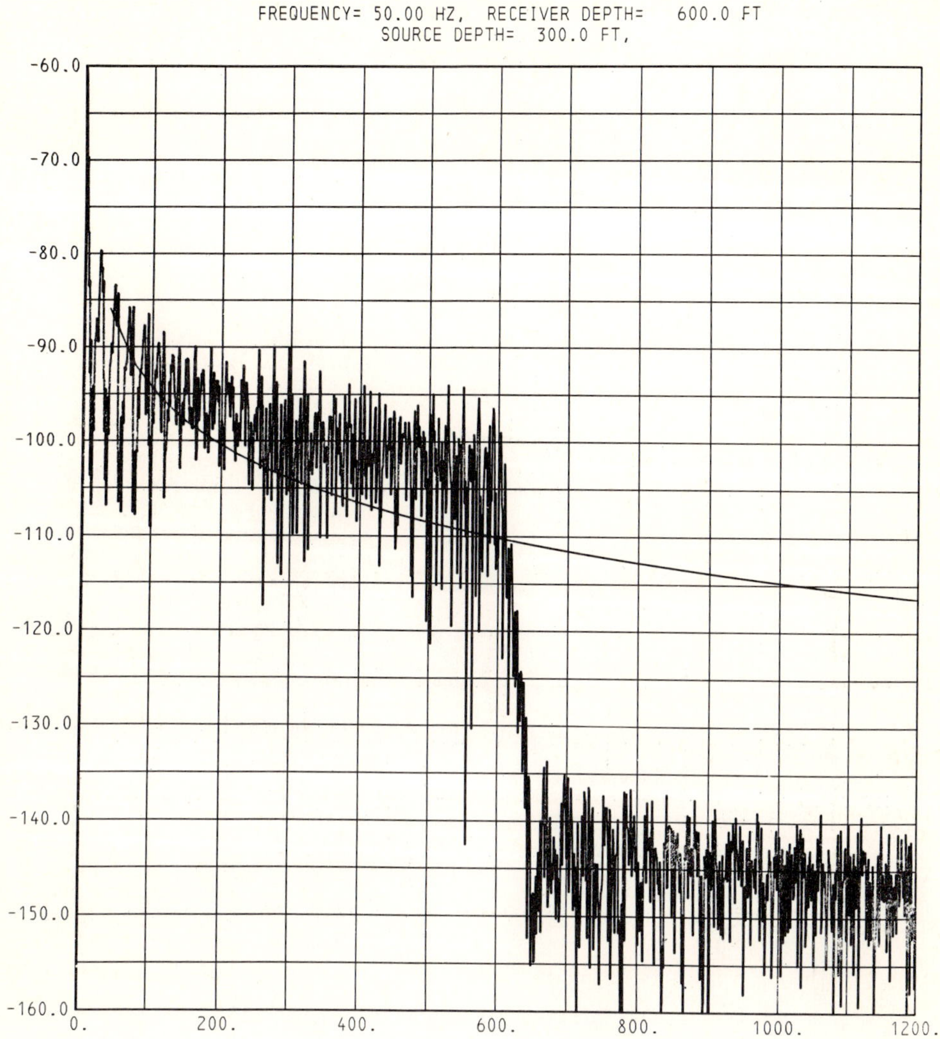

Figure 5. Predicted transmission loss across an oceanic front.

as shown in the numerically computed transmission loss curve of Fig. 5 for a source at 300 ft. and frequency 50 Hz. The cause of this drop in transmission is clearly the "impedance mismatch" between the two different sound channels separated by the oceanic front.

Next, we present some graphical computer plots of acoustic fields that were made for their pedagogical (and possibly artistic) value. In the first sequence, Figs. 6 to 10, the sound speed profile was taken to be bilinear in order to make the ray tracing easy to perform and to display the structure of caustics, shadow zones, and convergence zones in an idealized case. The gradient of sound speed, $g = dc/dz$ (with z increasing downward from the surface at $z = 0$), is given by: $g = -\ .04\ \text{sec}^{-1}$ between 0 and 4000 ft. (axis of the sound channel); and $g = +\ .02\ \text{sec}^{-1}$ between 4000 ft. and 16000 ft. (bottom of the ocean, assumed flat). The bottom layers were taken to be very lossy to eliminate bottom bounce paths and simplify the interpretation of the acoustic fields. The source (at left edge in all plots) is 2000 ft. deep and the horizontal scale extends to 80 nmi in all plots. Fig. 6 is a ray diagram of this case which was made by R. L. Holford [64]. Cusped caustics can be seen at the convergence zones at the source depth (2000 ft.) and between the convergence zones at the reciprocal depth (8000 ft.) of the source. Sharply defined shadow zones are also seen. Solutions of the parabolic wave equation for the same case are depicted in Figs. 7 - 10 at frequencies of 25, 50, 100, and 200 Hz, respectively, The upper plot shows detailed contour levels of acoustic intensity ($|\psi|^2$) in the range-depth plane. The lower plot is a smoothed and simplified version of the upper one. It shows two contour levels of $|p|^2$ corresponding to transmission losses of 80 dB (re 1 yd) and 90 dB (re 1 yd). The regions where the transmission loss is greater than 90 dB or less than 80 dB are shaded, while the region between 80 and 90 dB is left white. These plots show clearly the complicated interference and diffraction effects that are fully described by the parabolic wave equation but absent in ordinary ray training. To extend the geometrical acoustic approximation to take into account all of these effects would not appear to be a very practical approach, although it must be admitted that at the higher frequencies (200 Hz) the isolated caustics occurring at the first convergence zone may be adequately handled by such methods in this idealized case.

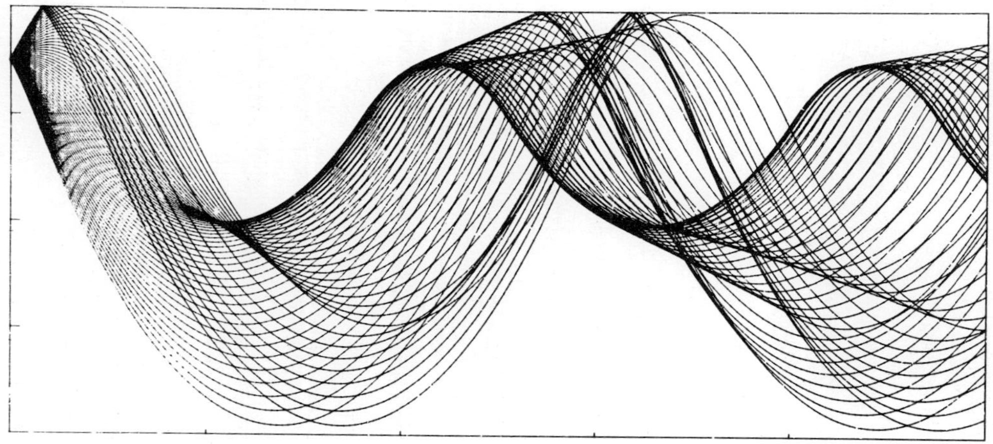

Figure 6. Bilinear profile, ray trace (by R. L. Holford).

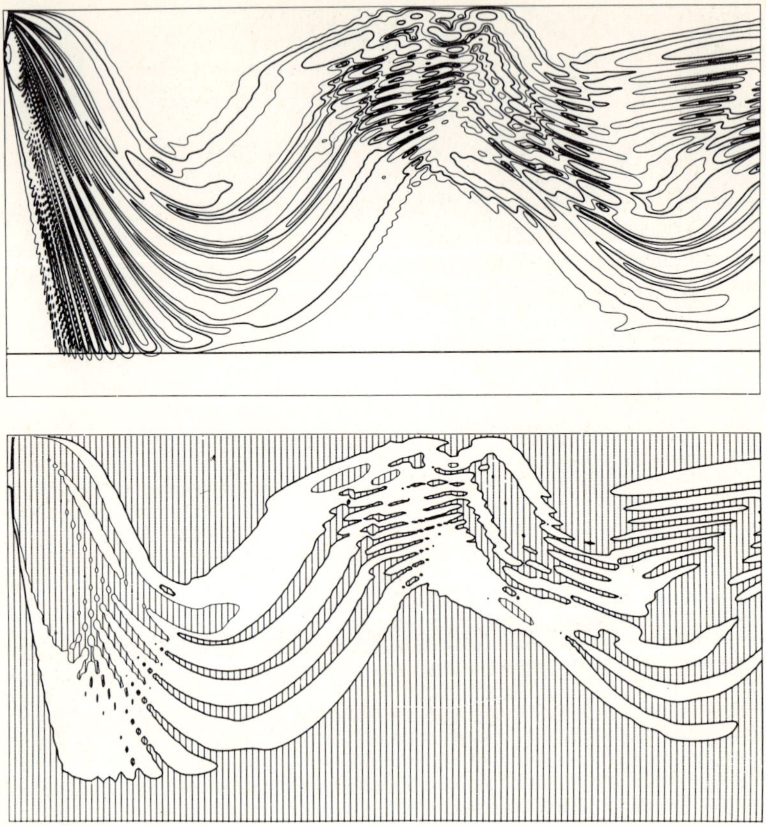

Figure 7. Bilinear profile, parabolic equation, 25 Hz.

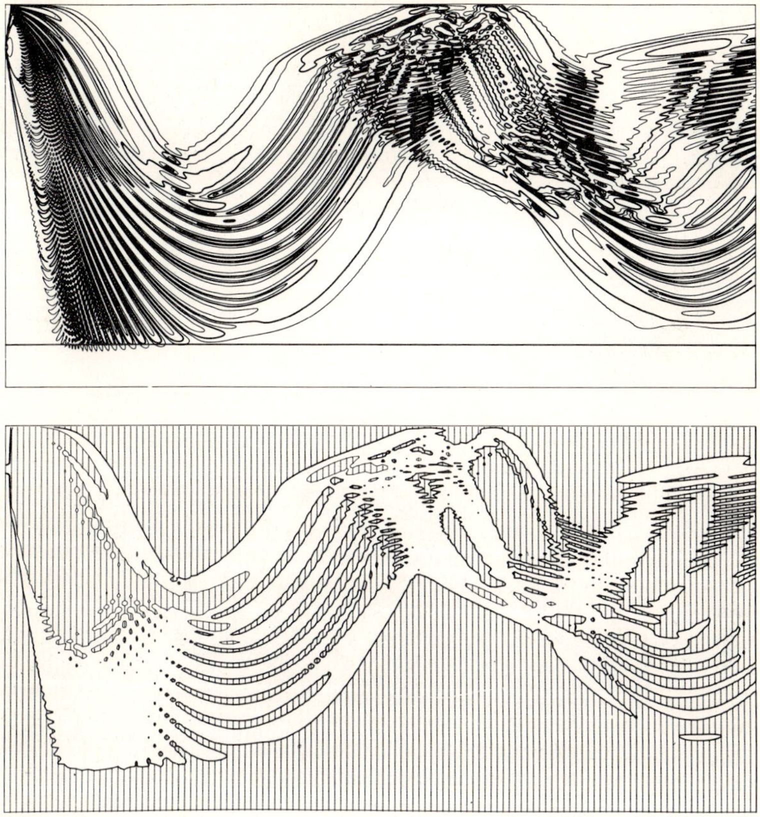

Figure 8. Bilinear profile, parabolic equation, 50 Hz.

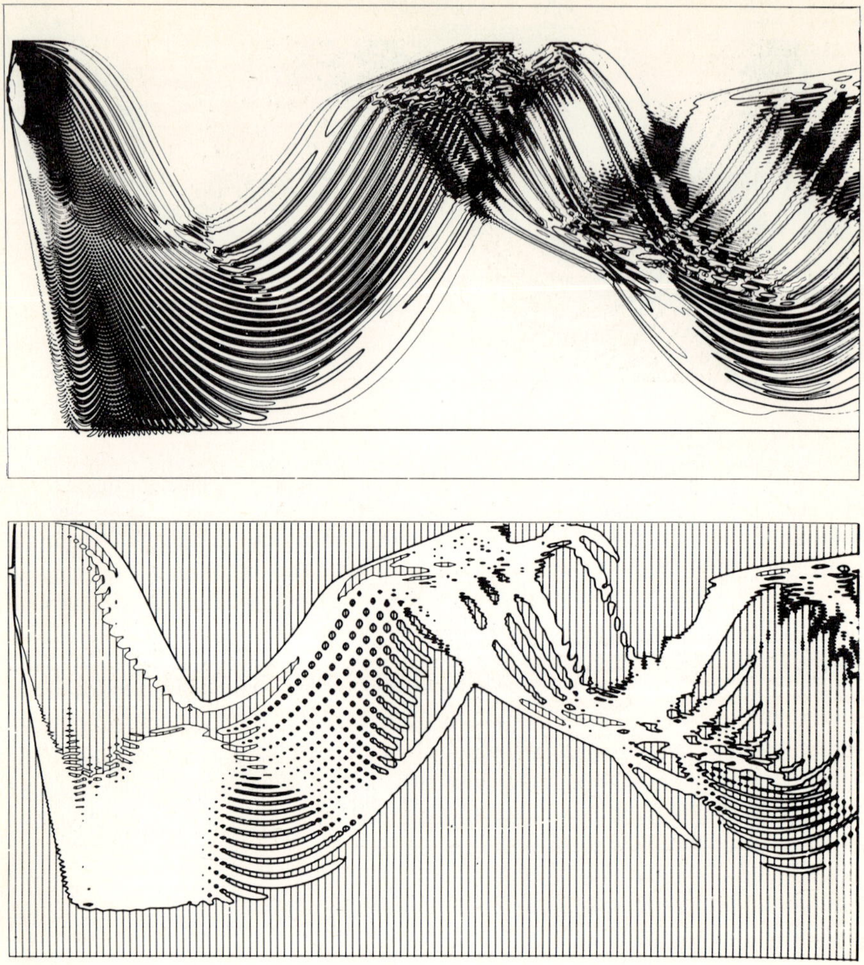

Figure 9. Bilinear profile, parabolic equation, 100 Hz.

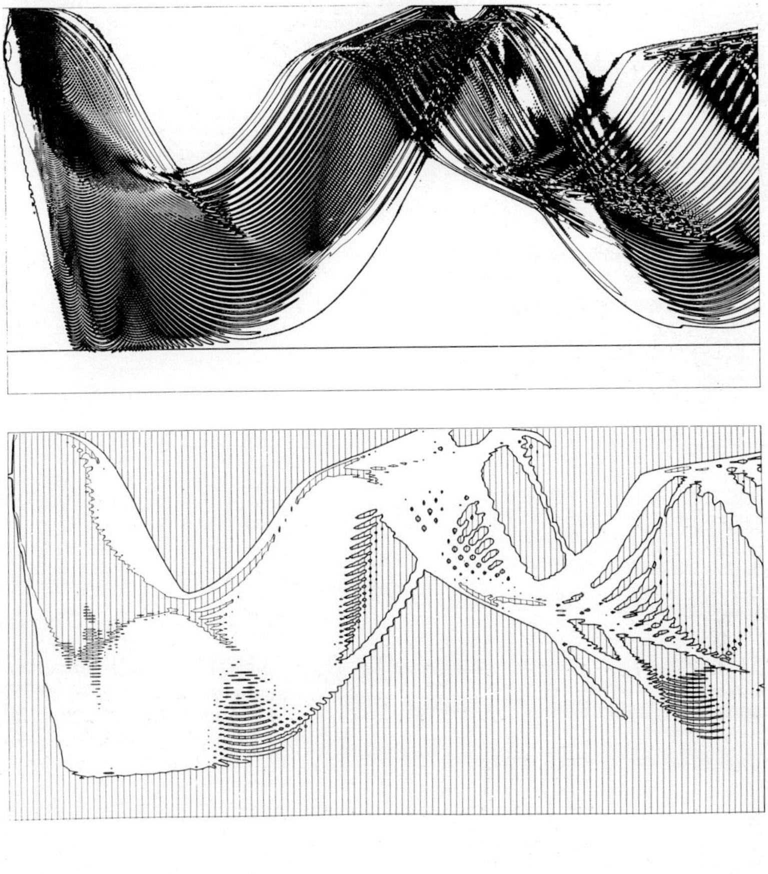

Figure 10. Bilinear profile, parabolic equation, 200 Hz.

Note, however, that at 25 Hz the caustic structures are so diffuse and overlapping that they are hardly recognizable. It is also instructive to note, while comparing Figs. 7 - 10, that the shadow zones are progressively being filled-in by diffraction effects at lower and lower frequencies. This, of course, is a well-known and experimentally observed effect.

Finally, we present three examples, Figs. 11 - 13, of acoustic propagation in the presence of an idealized sea-mount (or ridge), with W. Munk's canonical sound speed profile [42]:

2.18 $$c(z) = c_A[1 + \varepsilon(\eta - 1 + e^{-\eta})],$$

$$\eta = 2(z - z_A)/B \; .$$

In these examples, c_A = 1500 m/sec, ε = .0074, and z_A = B = 1.3 km. The floor of the ocean is at a depth of 4.5 km and the computational domain extends down to 5.0 km. The sea-mount rises halfway to the surface, or 2.25 km above the floor. The total horizontal range in all three examples is 400 km, and the frequency is 50 Hz. The plotting format is the same as in the previous examples except that the contour levels in the lower plot are 90 and 100 dB. In Figs. 11 and 12 the bottom (including the sea-mount) is again made very lossy so that all acoustic waves that interact with the bottom are completely attenuated. In Fig. 11 the source is 1.0 km deep, or slightly above the axis of the sound channel (at 1.3 km). It is seen that the effect of the sea-mount is to strip away the larger angle paths, leaving the near-axial paths virtually unaffected. This transmission down the axis is quite good, but not to a receiver near the surface (or vice versa). This is seen clearly in Fig. 12 where the source is now near the surface, 0.1 km deep. Propagation is by means of deep-cycling RSR paths, which are intercepted and absorbed by the sea-mount leaving almost no measurable acoustic signal. The last example, Fig. 13, is the same as in Fig. 12 except that the bottom is now not at all lossy, so that the reflection coefficient would be given by the Rayleigh formula. The critical angle in this case is about 10° at the ocean floor at 4.5 km and becomes slightly more on the sea-mount because the sound speed of the bottom material was held constant. One observes that

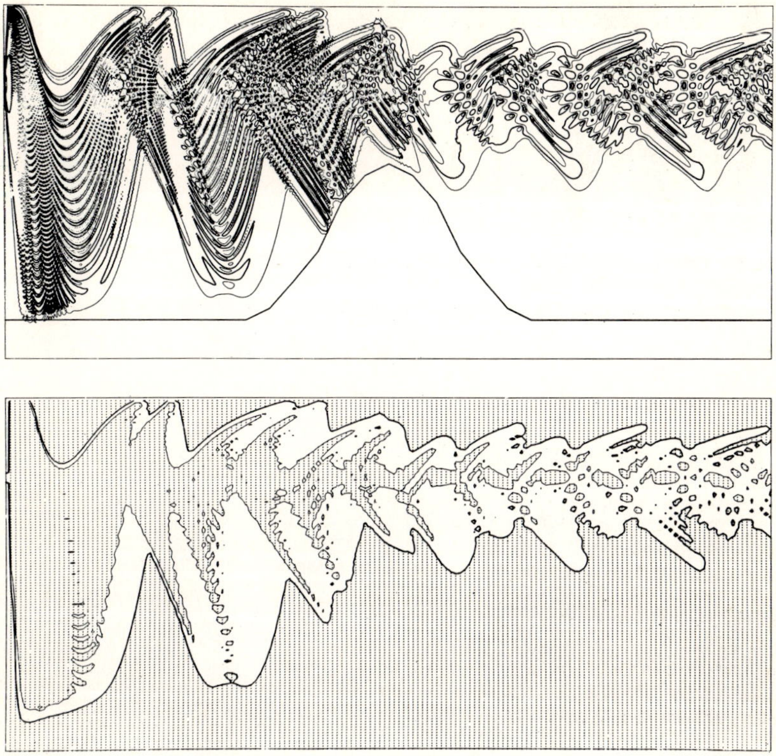

Figure 11. Canonical profile with sea mount, deep source, soft bottom.

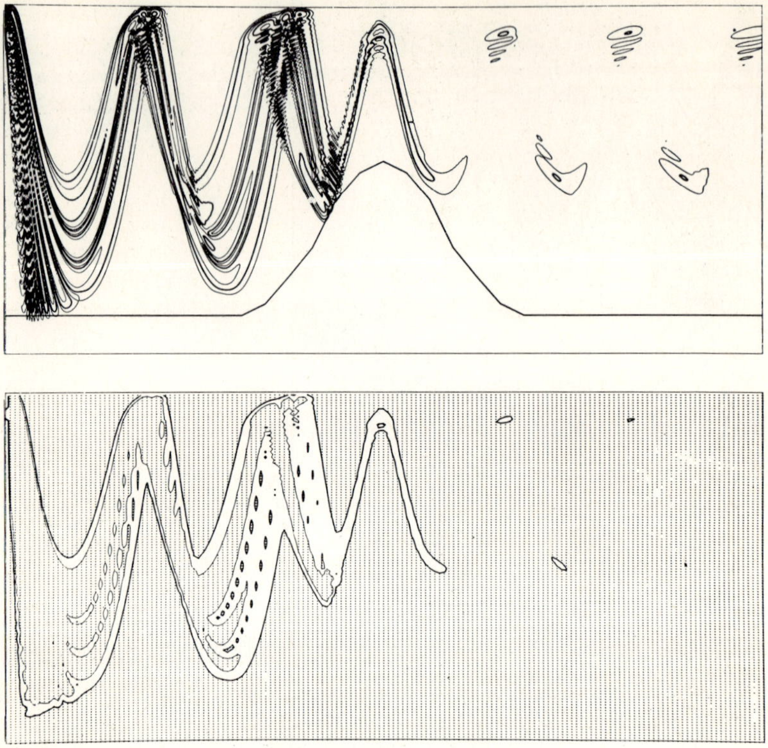

Figure 12. Canonical profile with sea mount, shallow source, soft bottom.

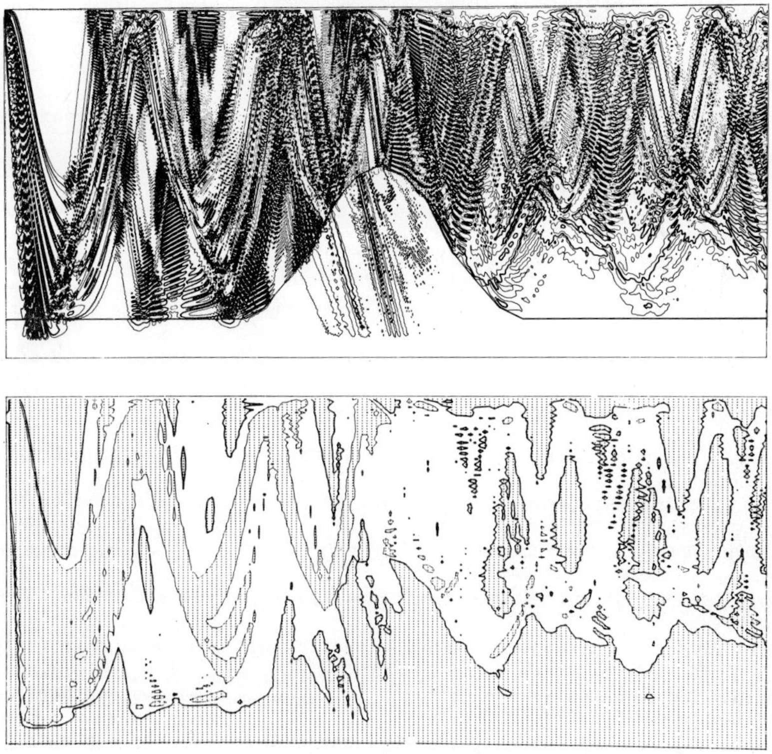

Figure 13. Canonical profile with sea mount, shallow source, hard bottom.

steep-angle waves refract and penetrate the bottom where they are eventually absorbed, whereas shallow-angle waves are totally reflected back into the water. The result is that transmission across the sea-mount is quite good, except that receivers near the bottom are screened by the sea-mount. One may also observe the sub-bottom region where the penetrating waves are artificially absorbed to prevent these waves from being reflected back into the water.

Numerical solutions of the parabolic wave equation produce in the course of solution full two dimensional acoustic fields; therefore contour plots of the above type require few additional calculations beyond those needed to solve the wave equation itself. When more realistic sound speed profiles, bottom depth profiles, and volume loss functions are used in such numerical calculations, one obtains as output not only transmission loss curves (such as shown in Fig. 4) whose numerical values can be compared directly to experimental data, but also plots of acoustic fields which give insight into the characteristics of acoustic wave propagation in realistic ocean environments. Thus one is in the enviable position of having a computer code that produces both insight <u>and</u> numbers, thereby confounding the aphorism that "the purpose of computing is insight, not numbers" [65].

Before ending this section, we shall discuss two additional effects -- variable density and earth curvature -- that were not included in the model described above but which can be added with little additional effort to gain a somewhat greater degree of realism.

In a fluid with given variable density ρ, the reduced wave equation for the acoustic pressure p is:

2.19 $$\rho\nabla\cdot\left(\frac{1}{\rho}\nabla p\right) + \frac{\omega^2}{c^2}\, p = 0,$$

where as before c is the variable sound speed. It is well-known [30,31] that the replacement,

2.20 $$q = p/\sqrt{\rho},$$

transforms eq. (2.19) into the standard Helmholtz form of the reduced wave equation:

2.21 $$\Delta q + k_o^2 n^2 q = 0,$$

where $k_o = \omega/c_o$, and the "effective" index of refraction is given by

2.22 $$n^2 = \frac{c_o^2}{c^2} + \frac{1}{2k_o^2}\left[\frac{1}{\rho}\Delta\rho - \frac{3}{2}\left(\frac{\nabla\rho}{\rho}\right)^2\right]$$

$$= \left(\frac{c_o}{c}\right)^2 + \frac{1}{2k_o^2}\rho^{1/2}\nabla\cdot(\rho^{-3/2}\nabla\rho).$$

The previous derivation of the parabolic wave equation can now be repeated (after inserting a volume loss term $i\nu$ which may now depend on density) by setting

2.23 $$q = \psi H_o^{(1)}(k_o r),$$

and obtaining

2.24 $$2ik_o\frac{\partial\psi}{\partial r} + \nabla^2\psi + k_o^2(n^2 - 1 + i\nu)\psi = 0,$$

which is exactly the same as (2.6) or (2.8) except that n is now given by (2.22).

A word of caution should be given concerning the use of this variable density model in cases where the density changes discontinuously. We see from (2.22) that large gradients in ρ will make the effective index of refraction change by large amounts, and yet the derivation of the parabolic wave equation requires that n^2 be nearly constant. In order to use this model in numerical simulations, one must therefore "smear out" the changes in density. However, to retain the correct scattering from density variations, one must not overdo this smearing. Consider, for example, a density profile that changes suddenly from ρ_1 to ρ_2 at the bottom of the ocean. A possible analytic expression for the density is

2.25 $$\rho(z) = \tfrac{1}{2}(\rho_1+\rho_2) + \tfrac{1}{2}(\rho_2-\rho_1)\tanh\left(\frac{z-H}{L}\right),$$

where L is the vertical distance over which the density changes. In order that the reflection from this density jump be correctly modeled, L must be small compared to the vertically projected wavelength, or

2.26 $$k_o^2 L^2 \sin^2 \theta \ll 1,$$

where θ is the angle of incidence (with respect to horizontal) and is usually very small. On the other hand, to avoid large values of n^2, (2.22) shows that L must be chosen such that

2.27 $$k_o^2 L^2 \gg \left| \frac{\rho_2-\rho_1}{\rho_2+\rho_1} \right| .$$

Both conditions can be satisfied only if $|\theta|$ is small and $|\rho_2-\rho_1|$ is not too large. Fortunately, this is the situation which commonly occurs in practice. For example, if $|\theta| \lesssim 10^o$ and $|(\rho_2-\rho_1)/(\rho_2+\rho_1)| \lesssim 1$, then the choice $k_o L = 2$ provides an adequate and useful approximation. One should not worry about making L depend on acoustic frequency because this model, if properly implemented, will ensure that the acoustic waves behave as though a discontinuity were present. It may also be worthwhile mentioning here that hydrophone sensors respond to the flux of acoustic energy which is equal to $|p|^2/\rho c = |q|^2/c$, so that computations of transmission loss with this model do not have to be renormalized with density ratios.

Lastly, we consider the effect of earth curvature on long-range acoustic propagation in the ocean. Letting r be the horizontal range from a source and R be radius of the earth, the mean level of the ocean surface as a function of range is

2.28 $$z_s = R - (R^2-r^2)^{1/2} \approx r^2/2R.$$

The sound speed profiles are measured downward from this surface, and the two-dimensional parabolic wave equation becomes

2.29 $$2ik_o \frac{\partial\psi}{\partial r} + \frac{\partial^2\psi}{\partial z^2} + k_o^2[n^2(r,z-z_s) - 1]\psi = 0.$$

We now make the transformation

2.30a $$z' = z - z_s = z - r^2/2R,$$

2.30b $$\psi(z,r) = \psi'(z',r)e^{ik_o r(z-r^2/3R)/R}.$$

In the transformed variables, the parabolic wave equation takes the form

2.31 $$2ik_o \frac{\partial\psi'}{\partial r} + \frac{\partial^2\psi'}{\partial z'^2} + k_o^2[n^2(r,z') - 1 - 2z'/R]\psi' = 0.$$

This transformation shows that, within the parabolic approximation, the effect of earth curvature is fully described by an additional term in the index of refraction which decreases linearly with depth. Thus the effective index of refraction is

2.32 $$n'^2 = n^2 - 2z'/R.$$

Assuming $n^2 \approx 1$, this may also be expressed as an effective sound speed:

2.33 $$c'(z',r) \approx c(z',r) + c_o z'/R.$$

The effective sound speed has an additional term increasing linearly with depth. Since the gradient of this additional term is $c_o/R \approx 2.5 \times 10^{-4}\ \mathrm{sec}^{-1}$, and since the gradient of c in the deep ocean has the nearly universal value $dc/dz \approx 1.7 \times 10^{-2}\ \mathrm{sec}^{-1}$, we see that the effect of earth curvature is nearly unmeasurable because it only modifies the usual deep ocean refraction by about 1%. Furthermore, the acoustic intensity is unchanged by the above transformation because $|\psi|^2 = |\psi'|^2$. Thus, although the effect is small it is easy to include earth curvature in the parabolic equation acoustic model.

In this section, we have developed a versatile acoustic model based on the parabolic wave equation and we have illustrated the use of the model in several numerical calculations. In the next section, we shall re-examine the range of validity of the model by formal asymptotic analysis.

3. Asymptotic Analysis

In this section, we shall present some formal asymptotic analysis of the parabolic approximation in the context of underwater acoustics in order to better understand its range of validity and possibly to increase its scope. We shall consider a two dimensional (r,z) model in which the far-field approximation has already been made. That is, we set $p(r,z) = u(r,z)/\sqrt{r}$ and assume $k_o r >> 1$ to obtain

3.1 $$u_{rr} + u_{zz} + k_o^2 n^2(z,r)u = 0.$$

As before, $k_o = \omega/c_o$, and $n = c_o/c(z,r)$ is the acoustic index with the dependence on azimuthal angle and time suppressed for notational convenience. We analyze this equation in two ways: first by means of a formal asymptotic scaling, and second by means of a factorization using pseudo-differential operators.

To use scaling arguments on (3.1), we introduce the dimensionless variables

3.2a $$z' = zk_o/f,$$

3.2b $$r' = rk_o/f^2,$$

where f is a dimensionless parameter which at present will be left unspecified. Later, we shall see that f is best identified with the f-number of the sound channel previously introduced in section 1. Defining the envelope function ψ by

3.3 $$u(z,r) = \psi(z',r') \exp(ik_o r),$$

and substituting this relation into (3.1), we obtain

3.4 $$\frac{1}{f^2}\psi_{r'r'} + 2i\psi_{r'} + \psi_{z'z'} + f^2(n^2-1)\psi = 0.$$

Next, we define the function η by

3.5 $$n^2 = 1 + \eta/f^2,$$

and consider the formal asymptotic behavior as $f \to \infty$. Ignoring for the moment the question of how η depends on z' and r', we make the expansion

3.6 $$\psi = \psi^{(0)} + \frac{1}{f^2}\psi^{(1)} + \frac{1}{f^4}\psi^{(2)} + \cdots ,$$

and equate powers of f^2 to get

3.7a $$2i\psi^{(0)}_{r'} + \psi^{(0)}_{z'z'} + \eta\psi^{(0)} = 0,$$

3.7b $$2i\psi^{(1)}_{r'} + \psi^{(1)}_{z'z'} + \eta\psi^{(1)} = -\psi^{(0)}_{r'r'} .$$

Eq. (3.7a) is the desired parabolic wave equation, and (3.7b) allows us to estimate the error made in neglecting the second, and higher, terms in the expansion (3.6). Assuming that at most $\psi^{(1)}$ grows linearly with r' [i.e., $\psi^{(1)} \sim r'\psi^{(0)}$], we see that the error will be order unity when $r'/f^2 \sim 1$. Thus we should restrict r to the range

3.8 $$r \lesssim f^4/k_o,$$

to ensure that the error is small.

We now must decide how to choose f. Three possible choices will be examined, and the third will be selected. The conventional choice is [19, 57]

3.9 $$f = k_oB,$$

where B is the width of the sound channel; i.e., the scale on which n varies with depth. This choice makes f extremely large and appears to give a very large range of validity according to (3.8). However, this f depends on frequency, and from (3.5) we see that in general we would have to require that

$$\eta = O(\Delta c/c_o) = O(1/f^2) = O(1/k_o^2 B^2). \tag{3.10}$$

Since $\Delta c/c_o$ is determined by oceanographic factors and is of course independent of frequency, we would have to restrict ourselves to essentially one frequency only, namely,

$$\nu = c_o k_o/2\pi \sim c_o \sqrt{c_o/\Delta c}\,/2\pi B \sim 2 \text{ Hz}.$$

We conclude that this choice is not suitable for general use. The severe restriction that we found here arose from the requirement that both the solution and the coefficient in the parabolic wave equation should vary on the same scale in z, namely B. However, as discussed in section 1 and as illustrated by numerical calculations in section 2, the focusing action of the sound channel causes the solution (the acoustic field) to vary on much shorter scales than the coefficient (the sound speed profile).

A second special case where both scales can be nicely balanced is the case of SOFAR propagation in a quadratic profile:

$$n^2 = 1 - \varepsilon(z/B)^2[1 + \delta c(z,r)/c_o], \tag{3.11}$$

$$\varepsilon = \Delta c/c_o \sim 10^{-2},$$

where the term $\delta c/c_o \ll 1$ accounts for deviations, possibly random, from the quadratic profile. In this case the appropriate choice of f is [20]

$$f = (k_o^2 B^2/\varepsilon)^{1/4}, \tag{3.12}$$

since (3.7a) becomes

3.13 $$2i\psi_{r'} + \psi_{z'z'} - z'^2[1 + \delta c/c_o]\psi = 0.$$

It may be noted that (3.8) becomes in this case

3.14 $$r \lesssim k_o B^2/\varepsilon \sim 10^4 \text{ km},$$

which agrees with the estimate given in Eq. (1.19). This is explained by the fact that the eigenvalue estimates given in section 1 were based on the lower modes of a quadratic profile (canonical for SOFAR propagation). We must now admit, however, that the above scaling and estimates are only valid for a very restricted class of problems and cannot be used for numerical examples presented in section 2 which clearly have quite different scales in the acoustic field and in the index of refraction.

We shall now show that the large parameter on which the parabolic approximation is based should be the f-number of the sould channel. Defining

3.15 $$n^2(z,r) = 1 + \varepsilon\eta(z,r),$$

$$\varepsilon = \Delta c/c_o \sim 10^{-2} \quad \text{(a fixed constant)},$$

$$\eta = O(1),$$

we conclude from (3.5) that the best choice for f for general applications in underwater acoustics is

3.16 $$f = 1/\sqrt{\varepsilon} \sim 10.$$

The same choice for the expansion parameter was motivated physically in section 1. The rms angle of propagation is $\theta \sim 1/f \ll 1$, and thus we again see that the parabolic approximation is a small angle approximation. We also note that the scaling in (3.2) with this choice of f re-affirms the statements made in section 1 about the scales on which the acoustic field varies in depth and range due to propagation in a continuous weakly focusing sound channel. The variation of $\psi^{(0)}$ with z' and r' is order unity if $\psi^{(0)}$ satisfies

3.17 $$2i\psi^{(0)}_{r'} + \psi^{(0)}_{z'z'} + \eta\psi^{(0)} = 0,$$

and η is assumed to vary with z on the scale B.

Expressing η in terms of z' and r', we find that

3.18 $$\eta(z,r) = \mu(z/B,\ r/A) = \mu\left(\frac{f}{k_o B} z',\ \frac{f^2}{k_o A} r'\right).$$

Thus in general, η will depend on the additional parameters, f/k_oB and f^2/k_oA, which cannot be allowed to become large (they may be small). This imposes the conditions

3.19 $$k_oB \gtrsim f = 1/\sqrt{\varepsilon}, \quad \text{or} \quad \nu \gtrsim 2\text{ Hz},$$

which gives the lower limit of validity of the parabolic approximation, and

3.20 $$A \gtrsim fB,$$

which determines the allowed rate of change of sound speed with range to be at least f times greater than its rate of change with depth. Finally, we note that on the basis of (3.8), the range of validity is limited to $r \lesssim 10^4/k_o$, or about 100 km at 100 Hz. This is a pessimistic estimate because we assumed the worst case, that $\psi^{(1)}$ grows linearly in r'. Assuming instead that errors tend to average out and that $\psi^{(1)} \sim \sqrt{r'}\,\psi^{(0)}$, we would instead conclude that

3.21 $$r \lesssim f^6/k_o,$$

on $r \lesssim 10^4$ km at 100 Hz. Experience has shown that this more optimistic estimate is probably closer to reality if one uses practical measures of accuracy such as average (over range) transmission loss, but that if one insists on absolute point-wise accuracy than the more pessimistic estimate is correct.

Next, we turn to another method of deriving parabolic wave equations which is based on "splitting" the solution of the elliptic wave equation (3.1) into a sum of two solutions: one propagating outward toward large r, and the other propagating inward toward small r (the backscatted, or reverberant wave). The preceeding method of derivation, based on asymptotic scaling and expansion, has the disadvantage that the correction terms must be successively smaller, order by order. Thus to get the first correction, $\psi^{(1)}/f^2$, one would solve (3.7a) for $\psi^{(0)}$, substitute this into the right hand side of (3.7b), and solve for $\psi^{(1)}$ by marching outward in r. If the resulting values of $\psi^{(1)}/f^2$ are small, one has gained very little (and besides, no backscatter effects are picked up). On the other hand, if $\psi^{(1)}/f^2$ is of order unity so that the correction is significant, then the higher order terms in (3.6) will also be order unity and the asymptotic expansion is no longer useful. The splitting method to be presented does not have this disadvantage, and it enables one to obtain corrections valid to all orders as well as a useful approximation for backscattered waves.

The starting point is again Eq. (3.1) which we here write in the form

3.22 $$\left(\frac{\partial^2}{\partial r^2} + k_o^2 Q^2\right)u = 0,$$

where the operator Q^2 is defined by

3.23 $$Q^2 = n^2(z,r) + \partial_z^2/k_o^2,$$

and $\partial_z^2 = \partial^2/\partial z^2$. In this and the next few paragraphs, we shall assume that the dependence of n on the range variable r is so weak (or absent) that we can neglect $\partial n/\partial r$ wherever it might appear. Later we shall return and pick up these neglected terms. Because of this assumption, the operator $\partial/\partial r$ commutes with Q^2, and we can formally factor (3.22) into two equations:

3.24a $$i \frac{\partial u_+}{\partial r} + k_o Q u_+ = 0,$$

3.24b $$-i \frac{\partial u_-}{\partial r} + k_o Q u_- = 0.$$

The full solution is the sum of the outgoing wave (u_+) and the incoming wave (u_-), i.e.,

3.25 $$u(z,r) = u_+(z,r) + u_-(z,r).$$

We note that in this approximation there is no coupling between u_+ and u_-. Thus if u_- vanishes initially, it will remain zero. Of course, when there is no range dependence in the index of refraction then this factorization is exact, and follows from the physically obvious fact that range variations of the ocean are necessary to couple outgoing and incoming waves.

In eqs. (3.24), Q is the pseudo-differential operator given formally by

3.26 $$Q = [n^2(z,r) + \partial_z^2/k_o^2]^{1/2}$$
$$= [1 + \varepsilon + \mu]^{1/2},$$

where

3.27 $$\varepsilon = n^2(z,r) - 1,$$

3.28 $$\mu = \partial_z^2/k_o^2.$$

We see that ε is a multiplication operator and μ is a differential operator (which happens to be second order with constant coefficient). The operator Q is called a pseudo-differential operator because, loosely speaking, it is a nonlocal operator, that is, $Qu(z)$ cannot be expressed in terms of a finite number of derivatives of u at the point z. The existing mathematical theory of such operators does not appear to extend to this particular example, chiefly because the radicand is not positive definite and thus a branch cut needs to be introduced into the definition of Q. Nevertheless, the proper way to do this is clear from the spectral decomposition of Q^2 (normal mode analysis): one chooses the exponentially decaying branch ($\mathrm{Im}\, Q > 0$) for the outgoing wave, and the opposite branch ($\mathrm{Im}\, Q < 0$) for the incoming wave. Thus, strictly speaking, we should use different expressions for Q in eqs. (3.24a) and (3.24b). However, this distinction disappears in the parabolic approximation, and we need not concern ourselves here with the evanescent modes.

The standard parabolic wave equation results from a truncation of the Taylor series expansion of the operator Q:

3.29 $$Q = 1 + \frac{1}{2}(\varepsilon+\mu) - \frac{1}{8}(\varepsilon+\mu)^2 + \cdots$$

Assuming that both ε and μ are small, we neglect the quadratic terms in (3.29) and substitute the remaining terms in (3.24a) to obtain

3.30 $$i\,\frac{\partial u}{\partial r} + k_o[1 + \frac{1}{2}(\varepsilon+\mu)]u = 0,$$

where the subscript on u has been omitted because we shall deal only with the outgoing wave in the next few paragraphs. Using the definitions of ε and μ given by (3.27) and (3.28) and making the usual envelope definition $u = \psi \exp(ik_o r)$, we obtain

3.31 $$i\,\frac{\partial\psi}{\partial r} + \frac{1}{2k_o}\,\partial_z^2\psi + \frac{k_o}{2}\,[n^2(z,r)-1]\psi = 0,$$

which is the usual parabolic wave equation.

We can now examine the conditions for validity of (3.31) from another point of view. We clearly need $||\varepsilon|| << 1$ and $||\mu|| << 1$ to make the local error small. As is often the case with approximations, estimates of cummulative errors are much more difficult to make and we shall not attempt to do so here. Now $\varepsilon = n^2-1 = c_o^2/c^2-1$ is determined by environmental conditions and by the choice of c_o. Since the sound speed $c(z)$ varies by only a few parts in a hundred through the water column, it is clear how to choose c_o so that $||\varepsilon||$ is small. For example, we may define

$$||\varepsilon|| = \int \varepsilon(z,r)|\psi|^2 dz / \int |\psi|^2 dz, \tag{3.32}$$

where ψ is the particular solution of (3.31) under consideration. This "norm" will depend on r, but it can easily be monitored during a numerical solution of (3.31) to see whether the parabolic approximation remains good. Since $|\psi|^2$ will tend to be large at a depth equal to the source depth (or the receiver depth in case one begins calculating from there), it follows from (3.32) that a good choice of c_o is the sound speed at the source depth since $\varepsilon = 0$ at this point and $||\varepsilon||$ will tend to be small. Numerical experience bears out this expectation.

Conditions under which μ is small are not so easy to state because, strictly speaking, μ is an unbounded operator. However, in the underwater acoustic applications we are only interested in the effect of μ acting on the acoustic field ψ at long ranges $(r >> B)$ and here ψ will vary slowly as a function of depth. Thus we define the "norm" of μ by

$$\begin{aligned} ||\mu|| &= \int \psi^* \mu \psi \, dz / \int |\psi|^2 dz \\ &= \int \left|\frac{1}{k_o} \frac{\partial \psi}{\partial z}\right|^2 dz / \int |\psi|^2 dz, \end{aligned} \tag{3.33}$$

where, as before, ψ is the particular solution of (3.31) under consideration. The physical meaning of $||\mu||$ is the mean square angle of propagation with respect to horizontal, since $\partial\psi/\partial z$ gives the vertical wavenumber and $k_o^{-1}\partial\psi/\partial z$ is the corresponding angle. This "norm" depends on r, and it too can easily be monitored during a numerical solution of (3.31), for example by using Parseval's relation and computing it in Fourier space. Therefore we have obtained an internal consistency check on the validity of the parabolic approximation: by monitoring the size of $||\varepsilon||$ and $||\mu||$, we can keep track of the relative errors made in the course of a calculation. This was in fact done in the numerical calculations described in section 2, and typically it was found that both $||\varepsilon||$ and $||\mu||$ remained less than about .04 throughout the calculations. It may also be mentioned that retaining higher order terms in the Taylor series expansion of Q given in (3.29) has the same disadvantage as with the asymptotic scaling expansion: when the correction terms are small they are not needed, and when they are large (order unity) then all higher terms must also be included.

We shall next present a derivation of improved parabolic equations which requires that only one of the operators (ε or μ) occurring in Q be small, but the other may be order unity. These equations represent significant improvements over the standard parabolic equation because they are valid to <u>all</u> orders in one of the operators ε or μ. Thus one is able to deal with index of refraction variations which are order unity in amplitude, $\varepsilon = O(1)$, or with propagation at large angles, $\mu = O(1)$. Of course if neither operator is small then one has no recourse except to return to the full elliptic equation. The basic idea in the derivation is the formal operator expansion

3.34a $$(A+\delta B)^{1/2} = A^{1/2} + \delta C + o(\delta^2),$$

3.34b $$C = \int_0^\infty e^{-A^{1/2}s} B e^{-A^{1/2}s} ds,$$

where A and B are operators (non-commuting, in general), and δ is a small constant. A formal proof of this relation is easily given by squaring both sides of (3.34a) to obtain the operator equation $B = A^{1/2}C + CA^{1/2}$, and noting that C as given by (3.34b) is a formal solution of this equation.

We now apply (3.34) to the operator Q given by (3.26). There are two cases: ε small or μ small. We first consider μ small. Neglecting terms of second order in μ, we obtain

3.36 $$\begin{aligned} Q &\approx (1+\varepsilon)^{1/2} + \int_0^\infty e^{-(1+\varepsilon)^{1/2}s} \mu e^{-(1+\varepsilon)^{1/2}s} ds \\ &= n + \int_0^\infty e^{-ns} (\partial_z^2/k_o^2) e^{-ns} ds. \end{aligned}$$

In this expression, $n = n(z,r)$ and is order unity. Although this operator appears formidable, it is actually easy to evaluate and is in fact a local operator. A straightforward calculation yields

$$\begin{aligned} Qu(z) &= nu + \frac{1}{k_o^2}\int_0^\infty e^{-2ns}(u_{zz} - sn_{zz}u - 2sn_z u_z + s^2 n_z^2 u)ds \\ &= nu + \frac{1}{2k_o^2}\left[\left(\frac{1}{n}u_z\right)_z + \frac{1}{2}\left(\frac{n_z^2}{n^3} - \frac{n_{zz}}{n^2}\right)u\right], \end{aligned} \tag{3.37}$$

where subscripts denote partial derivatives with respect to z. We then obtain a parabolic wave equation for u which is not substantially different from the standard version. Use of (3.37) in (3.24) gives

$$i\frac{\partial u}{\partial r} + \frac{1}{2k_o}\frac{\partial}{\partial z}\left(\frac{1}{n}\frac{\partial u}{\partial z}\right) + k_o\left[n + \frac{1}{4k_o^2}\left(\frac{n_z^2}{n^3} - \frac{n_{zz}}{n^2}\right)\right]u = 0. \tag{3.38}$$

This equation is valid to all orders in n^2-1, and has not previously been derived. To the author's knowledge, it has not yet been implemented numerically, although it would surely be worthwhile doing so. It may be noted that this equation is not equivalent to any of the modified parabolic equations that were obtained in Section 1 on the basis of replacing $(n^2-1)/2$ by $n-1$. The exact form of (3.38) would be difficult to guess by such means. It is also worth noting that this improved parabolic equation does not depend on the choice of the normalization sound speed c_o. This is because it may be written in such a way that k_o $(= \omega/c_o$ and n $(= c_o/c)$ always occur in the combination $k_o n$ $(= \omega/c)$ which is independent of c_o. This property must of course hold for any equation that is valid for all values of n. Further, (3.38) conserves the flux of outgoing acoustic radiation:

$$F_+ = \int |u|^2 dz = \text{const.}, \tag{3.39}$$

which is a reassuring fact.

Although (3.38) is a useful acoustic model as it stands, some numerical algorithms (such as the one described in [35-38, 49]) for solving such equations are effective only if the differential operator has constant coefficients, which is clearly not the case for (3.38). Therefore it is useful to transform (3.38) into an equivalent equation with constant coefficients. This can be done by changing the independent variable z to

3.40 $$\tilde{z} = \int^{z} [n(z')]^{1/2} dz',$$

and the dependent variable u to

3.41 $$\tilde{u} = [n(z)]^{-1/4} u.$$

Further, we define the index of refraction in terms of $\tilde{z}$ by

3.42 $$m(\tilde{z}) = n(z(\tilde{z})), \quad \text{or} \quad n(z) = m(\tilde{z}(z)).$$

A straightforward calculation then transforms (3.38) into

3.43 $$i\frac{\partial \tilde{u}}{\partial r} + \frac{1}{2k_o}\frac{\partial^2 \tilde{u}}{\partial \tilde{z}^2} + k_o[m(\tilde{z}) - \frac{1}{8k_o^2}(\frac{1}{m}\frac{\partial^2 m}{\partial \tilde{z}^2} + \frac{1}{4m^2}(\frac{\partial m}{\partial \tilde{z}})^2)]\tilde{u} = 0.$$

This equation has the desired form, and it conserves the energy flux since

3.44 $$F_+ = \int |u|^2 dz = \int |\tilde{u}|^2 d\tilde{z} = \text{const}.$$

Eq. (3.38), or its equivalent (3.43), is in a sense unique in that it is the only equation having the form of a parabolic wave equation which is correct to all orders in n^2-1. Other equations which do not have all the terms contained in (3.38) have not been systematically derived and cannot claim to be valid for large changes of the index of refraction.

Next, we examine the other way of expanding Q via (3.34). Assuming that ε is small but μ is order unity, we obtain

3.45 $$Q \approx (1+\mu)^{1/2} + \int_0^\infty e^{-(1+\mu)^{1/2}s}\, \varepsilon e^{-(1+\mu)^{1/2}s}\, ds$$

$$= (1+\partial_z^2/k_o^2)^{1/2} + \int_0^\infty e^{-(1+\partial_z^2/k_o^2)^{1/2}s}(n^2-1)e^{-(1+\partial_z^2/k_o^2)^{1/2}s}\, ds$$

Use of this expression in (3.24) gives

3.46 $$i\frac{\partial u}{\partial r} + (k_o^2+\partial_z^2)^{1/2}u + k_o^2\int_0^\infty ds e^{-(k_o^2+\partial_z^2)^{1/2}s}(n^2-1)e^{-(k_o^2+\partial_z^2)^{1/2}s}u = 0.$$

The integral operator occurring in this equation does not appear to be susceptible to further reduction. Nevertheless, it may be useful in numerical calculations where Fourier space methods are used to evaluate the exponential operators (bearing in mind the remarks made earlier about proper treatment of branch cuts). Eq. (3.46) is new, and it is the only improved parabolic wave equation known to the author which is valid for arbitrarily large angles (except for "exact" normal mode expansions). If, in the integral operator term of (3.46), one neglects ∂_z^2 compared to k_o^2 then this equation simplifies to

3.47 $$i\frac{\partial u}{\partial r} + (k_o^2 + \partial_z^2)^{1/2}u + \frac{k_o}{2}(n^2-1)u = 0.$$

Even though (3.47) is exact for propagation in an isovelocity ocean ($n^2 = 1$), it is not a systematic asymptotic equation for the general case because there is no justification for dropping ∂_z^2 in the integral operator and retaining it in the other term. Thus (3.47) is not a genuine improvement over the standard parabolic wave equation and its use should not be encouraged. This negative judgement about (3.47) is supported by numerical experience.

In the final portion of this section, we return to (3.22) and attempt to include the range dependence of n in the factorization of u into outward and inward propagating waves [25-29]. The main idea of this analysis is to simply transcribe the work of Bremmer on the second order ordinary differential equation analogous to (3.22) to the partial differential equation under consideration:

3.48 $$[\frac{\partial^2}{\partial r^2} + k_o^2 Q^2(r)]u = 0,$$

where $Q(r)$ is the operator defined previously in (3.23). Instead of (3.25) used before, we now split u according to

3.49a $$u = Q^{-1/2}(u_+ + u_-),$$

3.49b $$\frac{\partial u}{\partial r} = ik_o Q^{1/2}(u_+ - u_-).$$

These equations define u_+ and u_-, and we obtain

3.50a $$u_+ = \frac{1}{2} Q^{1/2}(u - \frac{i}{k_o Q} \frac{\partial u}{\partial r}),$$

3.50b $$u_- = \frac{1}{2} Q^{1/2}(u + \frac{i}{k_o Q} \frac{\partial u}{\partial r}).$$

It is now a simple matter to find the equations satisfied by u_+ and u_- by differentiating (3.50) with respect to r, using (3.48) for $\partial^2 u/\partial r^2$, and using (3.49) to replace u and $\partial u/\partial r$ in terms of u_+ and u_-. The resulting pair of coupled equations for u_+ and u_- are rather complicated and we shall not write them down here. When Q does not depend on r, they decouple and reduce to (3.24) previously derived. This fact demonstrates the main advantage of the factorization defined by (3.49), namely, that the equations for $u_\pm$ decouple exactly when there is no range dependence in the index of refraction. The price which one pays for maintaining this physical requirement is that one must deal with the nonlocal operator Q, fractional powers of Q, and commutators of the type $Q \cdot \partial Q/\partial r \pm \partial Q/\partial r \cdot Q$ [29].

In the following, we shall deal only with first order backscatter effects within the standard parabolic approximation. Thus we again use the Taylor series expansion of Q given in (3.29) and neglect quadratic terms. The result of this calculation is

3.51a $$i \frac{\partial u_+}{\partial r} + k_o[1 + \frac{1}{2}(\varepsilon+\mu)]u_+ = \frac{i}{4} \frac{\partial \varepsilon}{\partial r} u_-,$$

3.51b $$-i\frac{\partial u_-}{\partial r} + k_o[1 + \frac{1}{2}(\varepsilon+\mu)]u_- = -\frac{i}{4}\frac{\partial\varepsilon}{\partial r}u_+.$$

As remarked above, these equations are coupled only through the r dependence of $\varepsilon(z,r) = n^2(z,r)-1$. These equations may be used to compute acoustic reverberation as follows. We set

3.52 $$u_\pm = \psi_\pm e^{\pm ik_o r},$$

and neglect the twice scattered waves to obtain

3.53a $$i\frac{\partial\psi_+}{\partial r} + \frac{1}{2k_o}\frac{\partial^2\psi_+}{\partial z^2} + \frac{k_o}{2}[n^2(z,r)-1]\psi_+ = 0$$

3.53b $$-i\frac{\partial\psi_-}{\partial r} + \frac{1}{2k_o}\frac{\partial^2\psi_-}{\partial z^2} + \frac{k_o}{2}[n^2(z,r)-1]\psi_- = -\frac{i}{4}\frac{\partial n^2}{\partial r}\psi_+ e^{2ik_o r}.$$

Eq. (3.53a) is solved in the usual way for the outgoing wave by starting from $r = 0$ and marching out to the largest desired range. This stored solution is then put into the right hand side of (3.53b) and the solution of this equation is obtained by marching inward from large r backward toward the source at $r = 0$. In this manner, the acoustic energy scattered back from the environment to the source can be computed within the parabolic approximation. In principle, this procedure could be iterated: by sweeping forward and backward successively, one would build up the full solution of (3.48). In practice, the single-scatter approximation described above appears to be adequate.

4. Summary

This article has dealt with various aspects of parabolic approximation methods in underwater acoustics, mostly for propagation of sinusoidal signals. Extensions of these methods to time-dependent problems are also available: pulse propagation, moving sources and receivers, frequency shifting effects due to rapid temporal variations of oceanic conditions, and so forth. However, an adequate description of these extensions would require another long section and it was felt that the principles involved in making parabolic approximations have been sufficiently illustrated. Parabolic equation methods in underwater acoustics were developed only in the last few years, and as more and more use is made of these methods we may expect that many of the important modelling problems in ocean acoustics may be solved.

5. Acknowledgment

Most of the ideas in this article were first worked out in collaboration with Ronald H. Hardin in the years 1972-1973 while we were employed by Bell Laboratories, Inc. With pleasure, I acknowledge many useful discussions with this colleague.

Appendix A. Historical survey of parabolic wave equation applications

The "standard" parabolic wave equation in underwater acoustics has the same form as Schroedinger's equation in quantum mechanics, and thus mathematical studies of this equation go back to at least the mid-1920's. Indeed, this analogy provides a convenient point of entry for physicists going into underwater acoustics. However, as an approximation method in the theory of wave propagation, the parabolic wave equation dates from the work of Leontovich and Fock [1,2] in the mid-1940's. In fact, it was these scientists who coined the name "parabolic equation method". They applied the method to the problem of tropospheric radio wave propgation to long range (over the horizon). They were concerned with calculating the diffraction caused by the spherical shape of the earth, and the "preferred" direction needed to make the small-angle parabolic approximation was the line of sight between the antenna and the horizon. This method was later applied to many other radio wave diffraction problems [2,3] such as high frequency scattering by obstacles of various shapes. It has also been extensively applied [4,5] to the theory of microwave resonators, waveguides, and antennas.

When coherent sources of optical radiation (lasers) were developed in the early 1960's, it was a natural development to apply the parabolic equation method to problems of laser beam propagation, and this was quickly done [6,7]. In this field, the parabolic wave equation is usually called the "quasi-optical" equation. This equation is especially used for problems in nonlinear optics where the index of refraction depends on the intensity, thereby giving a nonlinear parabolic wave equation which is sometimes called the "nonlinear Schroedinger" equation. The parabolic equation method has also been applied to nonlinear optical pulse propagation in dielectric fibers [8], an area which is currently of great interest. In the past decade hundreds of research papers have been published by workers in nonlinear optics who use the "quasi-optical" approximation, and this research has recently been thoroughly reviewed [9,10].

In the field of plasma physics, there has occurred in recent years an enormous increase of interest in parabolic equation methods. Many types of waves can propagate in plasmas, and most work is concerned with nonlinear effects which in plasmas are especially large and significant. Some examples of such applications can be found in [11-14].

The parabolic equation method has also been extensively used since about 1968 to study the abstract problem of beam propagation in random media. The beams may consist of radio waves (radars), acoustic waves (sonars), optical waves (lasers), and so forth. This abstract problem is equivalent to the quantum mechanical problem of the motion of a particle in a random potential, and has been investigated by many scientists and applied mathematicians using a variety of techniques [15-21]. A concrete application of this method to the problem of radar beam propagation through randomly fluctuating ionospheres, including numerical simulations in three dimensions using the "split-step Fourier" algorithm, is given in [22].

In the field of seismic wave propagation, the parabolic equation method has been used since about 1970 [23] with no apparent awareness of its many other applications. These geophysical applications have been successful, and are thoroughly reviewed in [24].

The most recent application of the parabolic equation method to a concrete physical problem has been the subject of this article: low-frequency long-range underwater acoustic propagation. The early results of this application were reported in 1973-1974 [35-38]. A computer program was constructed which solves the parabolic wave equation using the split-step Fourier algorithm and accepts as input data measured oceanographic sound speed profiles and volume loss profiles (as functions of both range and depth) and ocean depth contours from nautical charts. Output data from the program (acoustic fields and transmission loss curves) were compared to experimental measurements and to other acoustic models with generally excellent results. At the same time, most of the theoretical considerations discussed in the main text of this article were developed and reported [35-38].

Interest is this new method spread rapidly, and soon groups of scientists at other laboratories developed their own computer programs based on the parabolic equation method, and the same split-step Fourier algorithm. The method was extended [39,43] to much higher acoustic frequencies than were originally contemplated with equally good numerical results. Another extension [40] to include random internal wave fluctuations [41] in the index of refraction was also quite successful. Since then, numerous additional investigations [44-57] of parabolic equation methods in underwater acoustics have been carried out at many laboratories, and this method is now (1976) widely available and routinely used for acoustic prediction studies*. The best available computer program for general acoustic use is called PE (for Parabolic Equation), and was developed at the Acoustic Environmental Support Detachment, Maury Center, Office of Naval Research [49,50].* Currently (1976), several groups are developing parabolic equation acoustic models that are three-dimensional and/or fully time-dependent.

---

*) These remarks (and references) were added much later than the lecture (on which this article is based) was presented.

References

[1] M. Leontovich and V. Fock, Zh. Eks. Teor. Fix. 16 (1946), 557-573 [J. Phys. USSR 10 (1946), 13-24; also Ref. [2], Chap. 11 .]

[2] V. A. Fock, Electromagnetic Diffraction and Propagation Problems, Pergamon Press, 1965 [Chaps. 11, 13, 14].

[3] G. D. Malyuzhinets, Usp. Fiz. Nauk 69 (1959), 321-334 [Sov. Phys. Usp. 2 (1959), 749-758].

[4] L. A. Vainshtein, Zh. Tekhn. Fiz. 34 (1964), 193 [Sov. Phys. Techn. Phys. 9 (1964), 157].

[5] L. A. Vainshtein, Open Resonators and Waveguides, Golem Press, Boulder, Colorado, 1969.

[6] P. L. Kelley, Phys. Rev. Lett. 15 (1965), 1005.

[7] V. I. Talanov, Zh. E. T. Fiz. Pis. Red. 2 (1965), 218 [JETP lett. 2 (1965), 138].

[8] A. Hasegawa and F. D. Tappert, Appl. Phys. Lett. 23 (1973), 142-144; 23 (1973), 171-172.

[9] O. Svelto, in Progress in Optics, Vol. 12 (ed. E. Wolf), North-Holland Pub., 1974 [Chap. I, pp. 1-51].

[10] J. H. Marburger, Prog. Quant. Elect. 4 (1975), 35-110.

[11] V. I. Karpman, Nonlinear Waves in Dispersive Media, Pergamon Press, 1975.

[11a] V. E. Zakharov, Zh. Eksp. Teor. Fiz. 62 (1972), 1745-1759 [Sov. Phys. JETP 35 (1972), 908-914].

[12] G. Morales and Y. C. Lee, Phys. Rev. Lett. 33 (1974), 1016-1019.

[13] H. H. Chen and C. S. Liu, Phys. Rev. Lett. 37 (1976), 693-697

[14] F. D. Tappert and C. N. Judice, Phys. Rev. Lett. 29 (1972), 1308-1311.

[15] V. I. Tatarskii, Zh. Eksp. Teor. Fiz. 56 (1969), 2106 [Sov. Phys. JETP 29 (1969(, 1133].

[16] L. A. Chernov, Sov. Phys. Acoust. 15 (1970), 511.

[17] V. I. Klyatskin and V. I. Tatarskii, Zh. Eksp. Teor. Fiz. 58 (1970), 624-634 [Sov. Phys. JETP 31 (1970), 335-339].

[18] Yu. N. Barabanenkov, Yu. A. Kravtsov, S. N. Rytov, and V. I. Tatarskii, Sov. Phys. Uspekhi 13 (1971), 551-580.

[19] P. L. Chow, J. Math. Phys. 13 (1972), 1224-1236.

[20] G. C. Papanicolaou, D. McLaughlin, and R. Burridge, J. Math. Phys. 14 (1973), 84-89.

[21] I. M. Besieris and F. D. Tappert, J. Math. Phys. 14 (1973), 1829-1836.

[22] R. H. Hardin and F. D. Tappert, "Analysis, simulation, and models of ionospheric scintillation", Joint Radar Propagation Study, Bell Lab Report, March 1974, 76pp.

[23] J. F. Claerbout, Geophys. 35 (1970), 407-418.

[24] J. F. Claerbout, "Numerical holography", in Acoustical Holography, Vol. 3, ed. A. F. Metherell, Plenum Press, N. Y., 1971, pp. 273-283.

[24a] J. F. Claerbout, Fundamentals of Geophysical Data Processing, McGraw-Hill, N N.Y., 1976.

[25] H. Bremmer, Comm. Pure Appl. Math. 4 (1951), 105-115.

[26] H. Bremmer, Radio Sci. 8 (1973), 511-534.

[27] P.v.d. Woude and H. Bremmer, Radio Sci. 10 (1975), 23-28.

[28] F. W. Sluijter, J. Opt. Soc. Am. 60 (1970), 8.

[29] J. Corones, J. Math. Anal. Appl. 50 (1975), 361-372.

[30] P. G. Bergman, J. Acoust. Soc. Am. 17 (1946), 329.

[31] L. M. Brekhovskikh, Waves in Layered Media, Academic Press, N. Y., 1970.

[32] I. Tolstoy and C. S. Clay, Ocean Acoustics, McGraw-Hill, N. Y., 1966.

[33] R. J. Urick, Principles of Underwater Sound, McGraw-Hill, N. Y., 1975.

[34] P. R. Tatro and C. W. Spofford, Proc. IEEE Ocean Environment Conf. (1973), pp. 206-216.

[35] F. D. Tappert and R. H. Hardin, in "A synopsis of the AESD workshop on acoustic modeling by non-ray tracing techniques", C. W. Spofford, AESD TN-73-05, Arlington, Va. (1973).

[36] R. H. Hardin and F. D. Tappert, SIAM Rev. (Chronicle) 15 (1973), 423; F. D. Tappert, SIAM Rev. (Chronicle) 16 (1974), 140.

[37] F. D. Tappert, J. Acoust. Soc. Am. 55 (1974), 534.

[38] F. D. Tappert and R. H. Hardin, Proc. Eighth Inter. Congress on Acoustics (London, 1974), Vol. 2, p. 452.

[39] G. W. Benthien, D. F. Gordon, and L. E. McCleary, J. Acoust. Soc. Am. 55 (1974), 545.

[40] S. M. Flatté and F. D. Tappert, J. Acoust. Soc. Am. 58 (1975), 1151-1159.

[41] C. Garrett and W. H. Munk, Geophys. Fluid Dyn. 2 (1972), 225-264.

[42] W. H. Munk, J. Acoust. Soc. Amer. 55 (1974), 220-226.

[43] S. T. McDaniel, J. Acoust. Soc. Amer. 57 (1975), 307-311.

[44] S. T. McDaniel, J. Acoust. Soc. Amer. 58 (1975), 1178-1185.

[45] A. O. Williams, Jr., J. Acoust. Soc. Amer. 58 (1975), 1320-1321.

[46] R. M. Fitzgerald, J. Acoust. Soc. Amer. 57 (1975), 839-842.

[47] E. A. Polyanskii, Sov. Phys. Acoust. 20 (1974), 90.

[48] J. A. DeSanto, "Connection between the solutions of the Helmholtz and parabolic equations for sound propagation", Proc. SACLANTCEN Conf. (1975).

[49] H. K. Brock, "The AESD parabolic equation model", AESD TN-75-07, ONR, Arlington, Va. (1975).

[50] H. K. Brock, R. N. Buchal, and C. W. Spofford, "Modifying the sound speed profile to improve the accuracy of the parabolic equation technique", AESD Tech. Memo (1975).

[51] D. R. Palmer, J. Acoust. Soc. Am. 60 (1976), 343-354.

[52] J. S. Hanna, J. Acoust. Soc. Am. 60 (1976), 1024-1031.

[53] R. N. Buchal and F. D. Tappert, "A variable range step in the split-step Fourier algorithm", AESD Tech. Memo (1975).

[54] F. Jensen and H. Krol, "The use of the parabolic equation method in sound propagation modelling", SACLANTCEN Memo SM-72 (1975).

[55] P. M. Volk, "Solutions of acoustic wave propagation in the ocean by the parabolic approximation", Thesis, Univ. of Hawaii, Dept. of Oceanography (1975).

[56] D. J. Thomson, "Parabolic equation acoustic model", Defense Research Establishment Pacific Report (1975).

[57] G. A. Kreigsmann and E. N. Larsen, "On the parabolic approximation to the reduced wave equation", SIAM J. Appl. Math. (to appear).

[58] H. F. Harmuth, J. Math. Phys. (MIT) 36 (1957), 269-278.

[59] A. V. Popov, Akust. Zh. 15 (1969), 265-274 [Sov. Phys. Acoust. 15 (1969), 226-233].

[60] A. L. Dyshko, Zh. vychisl. Mat. mat. Fiz. 8 (1968), 238-242 [USSR Comput. Math. Phys. 8 (1968), 340-346].

[61] A. Goldberg, H. M. Schey, and J. L. Schwartz, Amer. J. Phys. 35 (1967), 177.

[62] R. D. Richtmyer and K. W. Morton, Difference Methods for Initial-Value Problems, Second Ed., Interscience Pub., N. Y., 1967.

[63] T. Talpay, Bell Laboratories Report, 1972 (unpublished).

[64] R. L. Holford and C. W. Spofford, Bell Laboratories Report, Long Range Acoustic Propagation Project, April 1973 (unpublished).

[65] R. W. Hamming, Numerical Methods for Scientists and Engineers, 2nd Ed., McGraw-Hill, N.Y., 1973.

Topics in Applied Physics

Founded by **Helmut K. V. Lotsch**

This book series is devoted to research achievements of current interest. Each volume deals with a different topic under the editorship of a recognized authority in the field. It covers application-oriented aspects of the topic under consideration, the basic physical principles being summarized in a comprehensive introduction.
The contributors to each volume are internationally known experts. The publication periods are comparable with those of scientific journals to keep pace with the rapidly accumulating results.

Springer-Verlag
Berlin
Heidelberg
New York

Volume 1
Dye Lasers
Editor: **F.P. Schäfer**
114 figures. XI, 285 pages. 1973

Volume 2
Laser Spectroscopy
of Atoms and Molecules
Editor: **H. Walther**
137 figures, 22 tables
XVI, 383 pages. 1976
ISBN 3-540-07324-8

Volume 3
Numerical and Asymptotic Techniques in Electromagnetics
Editor: **R. Mittra**
112 figures. XI, 260 pages. 1975
ISBN 3-540-07072-9

Volume 4
Interactions on Metal Surfaces
Editor: **R. Gomer**
112 figures. XI, 310 pages. 1975
ISBN 3-540-07094-X

Volume 5
Mössbauer Spectroscopy
Editor: **U. Gonser**
96 figures. XVIII, 241 pages. 1975
ISBN 3-540-07120-2

Volume 6
Picture Processing and Digital Filtering
Editor: **T.S. Huang**
113 figures. XIII, 289 pages. 1975
ISBN 3-540-07202-0

Volume 7
Integrated Optics
Editor: **T. Tamir**
99 figures. XIII, 315 pages. 1975
ISBN 3-540-07297-7

Volume 8
Light Scattering in Solids
Editor: **M. Cardona**
111 figures, 3 tables
XIII. 339 pages. 1975
ISBN 3-540-07354-X

Volume 9
Laser Speckle and Related Phenomena
Editor: **J.C. Dainty**
133 figures. XIII, 286 pages. 1975
ISBN 3-540-07498-8

Volume 10
Transient Electromagnetic Fields
Editor: **L.B. Felsen**
111 figures. XIII, 274 pages. 1976
ISBN 3-540-07553-4

Volume 11
Digital Picture Analysis
Editor: **A. Rosenfeld**
114 figures. 47 tables.
XIII, 351 pages. 1976
ISBN 3-540-07579-8

Volume 12
Turbulence
Editor: **P. Bradshaw**
47 figures, XI, 335 pages. 1976
ISBN 3-540-07705-7

Volume 13
High-Resolution Laser Spectroscopy
Editor: **K. Shimoda**
132 figures. XIII, 378 pages. 1976
ISBN 3-540-07719-7

Volume 14
Laser Monitoring of the Atmosphere
Editor: **E.D. Hinkley**
84 figures. XV, 380 pages. 1976
ISBN 3-540-07743-X

Volume 15
Radiationless Processes in Molecules and Condenses Phases
Editor: **F.K. Fong**
67 figures XIII, 360 pages. 1976
ISBN 3-540-07830-4

Volume 16
Nonlinear Infrared Generation
Editor: **Y.-R. Shen**
134 figures. XI, 279 pages. 1977
ISBN 3-540-07945-9

Selected Issues from

Lecture Notes in Mathematics

Vol. 431: Séminaire Bourbaki – vol. 1973/74. Exposés 436–452. IV, 347 pages. 1975.

Vol. 433: W. G. Faris, Self-Adjoint Operators. VII, 115 pages. 1975.

Vol. 434: P. Brenner, V. Thomée, and L. B. Wahlbin, Besov Spaces and Applications to Difference Methods for Initial Value Problems. II, 154 pages. 1975.

Vol. 440: R. K. Getoor, Markov Processes: Ray Processes and Right Processes. V, 118 pages. 1975.

Vol. 442: C. H. Wilcox, Scattering Theory for the d'Alembert Equation in Exterior Domains. III, 184 pages. 1975.

Vol. 446: Partial Differential Equations and Related Topics. Proceedings 1974. Edited by J. A. Goldstein. IV, 389 pages. 1975.

Vol. 448: Spectral Theory and Differential Equations. Proceedings 1974. Edited by W. N. Everitt. XII, 321 pages. 1975.

Vol. 449: Hyperfunctions and Theoretical Physics. Proceedings 1973. Edited by F. Pham. IV, 218 pages. 1975.

Vol. 458: P. Walters, Ergodic Theory – Introductory Lectures. VI, 198 pages. 1975.

Vol. 459: Fourier Integral Operators and Partial Differential Equations. Proceedings 1974. Edited by J. Chazarain. VI, 372 pages. 1975.

Vol. 461: Computational Mechanics. Proceedings 1974. Edited by J. T. Oden. VII, 328 pages. 1975.

Vol. 463: H.-H. Kuo, Gaussian Measures in Banach Spaces. VI, 224 pages. 1975.

Vol. 464: C. Rockland, Hypoellipticity and Eigenvalue Asymptotics. III, 171 pages. 1975.

Vol. 468: Dynamical Systems – Warwick 1974. Proceedings 1973/74. Edited by A. Manning. X, 405 pages. 1975.

Vol. 470: R. Bowen, Equilibrium States and the Ergodic Theory of Anosov Diffeomorphisms. III, 108 pages. 1975.

Vol. 474: Séminaire Pierre Lelong (Analyse) Année 1973/74. Edité par P. Lelong. VI, 182 pages. 1975.

Vol. 484: Differential Topology and Geometry. Proceedings 1974. Edited by G. P. Joubert, R. P. Moussu, and R. H. Roussarie. IX, 287 pages. 1975.

Vol. 487: H. M. Reimann und T. Rychener, Funktionen beschränkter mittlerer Oszillation. VI, 141 Seiten. 1975.

Vol. 489: J. Bair and R. Fourneau, Etude Géométrique des Espaces Vectoriels. Une Introduction. VII, 185 pages. 1975.

Vol. 490: The Geometry of Metric and Linear Spaces. Proceedings 1974. Edited by L. M. Kelly. X, 244 pages. 1975.

Vol. 503: Applications of Methods of Functional Analysis to Problems in Mechanics. Proceedings 1975. Edited by P. Germain and B. Nayroles. XIX, 531 pages. 1976.

Vol. 507: M. C. Reed, Abstract Non-Linear Wave Equations. VI, 128 pages. 1976.

Vol. 509: D. E. Blair, Contact Manifolds in Riemannian Geometry. VI, 146 pages. 1976.

Vol. 515: Bäcklund Transformations. Nashville, Tennessee 1974. Proceedings. Edited by R. M. Miura. VIII, 295 pages. 1976.

Vol. 516: M. L. Silverstein, Boundary Theory for Symmetric Markov Processes. XVI, 314 pages. 1976.

Vol. 518: Séminaire de Théorie du Potentiel, Proceedings Paris 1972–1974. Edité par F. Hirsch et G. Mokobodzki. VI, 275 pages. 1976.

Vol. 522: C. O. Bloom and N. D. Kazarinoff, Short Wave Radiation Problems in Inhomogeneous Media: Asymptotic Solutions. V. 104 pages. 1976.

Vol. 523: S. A. Albeverio and R. J. Høegh-Krohn, Mathematical Theory of Feynman Path Integrals. IV, 139 pages. 1976.

Vol. 524: Séminaire Pierre Lelong (Analyse) Année 1974/75. Edité par P. Lelong. V, 222 pages. 1976.

Vol. 525: Structural Stability, the Theory of Catastrophes, and Applications in the Sciences. Proceedings 1975. Edited by P. Hilton. VI, 408 pages. 1976.

Vol. 526: Probability in Banach Spaces. Proceedings 1975. Edited by A. Beck. VI, 290 pages. 1976.

Vol. 527: M. Denker, Ch. Grillenberger, and K. Sigmund, Ergodic Theory on Compact Spaces. IV, 360 pages. 1976.

Vol. 532: Théorie Ergodique. Proceedings 1973/1974. Edité par J.-P. Conze and M. S. Keane. VIII, 227 pages. 1976.

Vol. 538: G. Fischer, Complex Analytic Geometry. VII, 201 pages. 1976.

Vol. 543: Nonlinear Operators and the Calculus of Variations, Bruxelles 1975. Edited by J. P. Gossez, E. J. Lami Dozo, J. Mawhin, and L. Waelbroeck, VII, 237 pages. 1976.

Vol. 552: C. G. Gibson, K. Wirthmüller, A. A. du Plessis and E. J. N. Looijenga. Topological Stability of Smooth Mappings. V, 155 pages. 1976.

Vol. 556: Approximation Theory. Bonn 1976. Proceedings. Edited by R. Schaback and K. Scherer. VII, 466 pages. 1976.

Vol. 559: J.-P. Caubet, Le Mouvement Brownien Relativiste. IX, 212 pages. 1976.

Vol. 561: Function Theoretic Methods for Partial Differential Equations. Darmstadt 1976. Proceedings. Edited by V. E. Meister, N. Weck and W. L. Wendland. XVIII, 520 pages. 1976.

Vol. 564: Ordinary and Partial Differential Equations, Dundee 1976. Proceedings. Edited by W. N. Everitt and B. D. Sleeman. XVIII, 551 pages. 1976.

Vol. 565: Turbulence and Navier Stokes Equations. Proceedings 1975. Edited by R. Temam. IX, 194 pages. 1976.

Vol. 566: Empirical Distributions and Processes. Oberwolfach 1976. Proceedings. Edited by P. Gaenssler and P. Révész. VII, 146 pages. 1976.

Vol. 570: Differential Geometrical Methods in Mathematical Physics, Bonn 1975. Proceedings. Edited by K. Bleuler and A. Reetz. VIII, 576 pages. 1977.

Vol. 572: Sparse Matrix Techniques, Copenhagen 1976. Edited by V. A. Barker. V, 184 pages. 1977.